Regulatory Data Science for Medical Devices

Regulatory Data Science for Medical Devices

Editor

Jeroen Bergmann

MDPI • Basel • Beijing • Wuhan • Barcelona • Belgrade • Manchester • Tokyo • Cluj • Tianjin

Editor
Jeroen Bergmann
University of Oxford
UK

Editorial Office
MDPI
St. Alban-Anlage 66
4052 Basel, Switzerland

This is a reprint of articles from the Special Issue published online in the open access journal *Prosthesis* (ISSN 2673-1592) (available at: https://www.mdpi.com/journal/prosthesis/special_issues/medical_devices_prosthesis).

For citation purposes, cite each article independently as indicated on the article page online and as indicated below:

LastName, A.A.; LastName, B.B.; LastName, C.C. Article Title. *Journal Name* **Year**, *Volume Number*, Page Range.

ISBN 978-3-0365-3893-8 (Hbk)
ISBN 978-3-0365-3894-5 (PDF)

Contents

About the Editor

Jeroen Bergmann is Associate Professor in the Department of Engineering Science at the University of Oxford. He is the Director of the Oxford Healthtech Labs and Official Fellow at Reuben College. He has a keen interest in Biomedical Entrepreneurship and has worked on a range of industry collaborations and business development projects. He became interested in using data science methods to tackle issues with navigating medical device regulations. He is now applying quantitative methods to the field of medical regulations to create new insights in the field of Medical Regulations.

prosthesis MDPI

Editorial

The Emerging Field of Medical Regulatory Technology and Data Science

Jeroen H. M. Bergmann

Department of Engineering Science, University of Oxford, Parks Road, Oxford OX13PJ, UK; jeroen.bergmann@eng.ox.ac.uk

Citation: Bergmann, J.H.M. The Emerging Field of Medical Regulatory Technology and Data Science. *Prosthesis* **2022**, *4*, 169–171. https://doi.org/10.3390/prosthesis4020017

Received: 2 April 2022
Accepted: 6 April 2022
Published: 9 April 2022

Regulations contain rules setup by (governmental) authorities to control specific aspects of certain industries, which often influences the way companies operate. These rules affect how industries are managed, and the importance of regulations is such that many companies create specific divisions focused solely on the regulatory strategy. Regulatory frameworks encourage consumers to adopt innovations by ensuring that their safety and effectiveness has been evaluated. However, they also create barriers that can hold up the innovative process. For innovative firms, regulations are one of the most significant barriers of perceived environmental uncertainty [1], which is especially problematic for start-ups with constrained resources [2]. Entrepreneurs need more information to better identify the relevant regulation and understand the requirements for conformity based on these regulations [3], which are the initial steps in the regulatory navigation pathway. The exact impact of regulation on innovation varies between both industries and countries [2]. Certain sectors, such as finance, energy and medical products, are rigorously regulated. It has been suggested that the successful disruption of an industry from pioneering innovation is always followed by regulation. A strategy must be in place to react to the legislations introduced as a result of new innovation. An example of this is data protection laws, which resulted from an increased level of personal information being stored by organizations due to advances in medical care, telecommunications, transportation systems and financial transfers [4]. When used effectively, regulation drives the direction of innovation and can stimulate it within industries [5]. Environmental policies forced car manufacturers to improve gas mileage, resulting in improvements in engine technology. A better understanding of legislation can help to alleviate the barrier to innovation that regulation presents.

Technology and data science has become an integrated part of how many industries operate, and it often affects their regulatory strategy. The rapid expansion of digital technology has also started to impact regulations themselves. Not only is legal information now available in a digital form, but some of the data held by regulators have become freely available online. The particular intersection between regulations and technology is known as Regtech. The main focus of Regtech is to support the different processes that are related to regulations. RegTech was initially suggested for addressing regulatory challenges in the financial system, through the use of innovative technologies [6]. However, the term has evolved to capture any area of regulation, including medical regulation. Buckley et al. have stated that RegTech can help create more effective and efficient ways to comply with the regulations [6]. RegTech can be applied to obtain better regulatory compliance or give the same level of compliance at a lower-cost. It is easy to see that both these outcomes are valuable for those working in the medical technology sector.

As mentioned, regulations are a key part of the medical innovation roadmap. It provides a framework to ensure patient safety and aims to guarantee a beneficial clinical performance of novel solutions. Any medical device that wants to be brought onto a (regulated) market needs to adhere to the regulations that have been set out by governments. All major markets in the world are regulated, and thus manufactures need to think carefully about their regulatory strategy. At the moment, there is relatively little research on RegTech within the medical field. Nonetheless, this is very likely to change in the future, due to the potential benefits it can offer to the different stakeholders (such as manufacturers, regulators and healthcare professionals). The digitisation, data availability and need for more concise understanding of regulations are all driving this change. We have already seen a similar thing happening within the financial sector, where the label Fintech was coined for financial RegTech. However, FinTech encapsulates more than just regulatory technology, as it also covers the technology which replaced traditional financial services. The term FinTech is thus not limited to just regulations, and this has muddied the waters somewhat. Despite this limitation, value can still be derived from a common term that can capture relevant regulatory developments in a specific field.

Such a label is currently missing for medical regulations, whilst it could be captured by the introduction of a new term, such as "MedRegTech". Unfortunately, the term MedTech is not appropriate in this case, as it represents medical technology more generally. This issue mirrors the problem described with the label FinTech. Introducing a more specific term of MedRegTech should allow for an easier classification of (scientific) articles that apply or describe RegTech in the medical area. This will help the (research) community to find relevant studies quicker, whilst making it easier to identify new research trends.

One of the benefits of MedRegTech research is that it could inform policymakers in an objective and critical manner. For example, the exploration of complexity within the regulations can provide new insights into how regulations can be made more user-friendly. We have showed that complexity can vary between the different medical device regulations [7]. Creating less-complex regulations, without losing the legal context of the regulation, can increase overall adherence and understanding. Moreover, this research also found that there is a need for better metrics in terms of regulatory complexity in general. How we best define complexity within regulations is still an open question. These kinds of studies can therefore provide starting points in the debate of determining an appropriate level of regulatory complexity.

Understanding regulations is essential for medical innovators, as they will need to be able to navigate them. Developing new ways to help people navigate the regulations forms another interesting avenue of exploration. Decision trees that are rule-based can potentially help with this. They offer an approach to the digitisation of the regulation that is logic-based [8,9]. These techniques are not perfect, and a good understanding of the context is still needed in order to apply them correctly. Yet, at the same time, they can also bring to the surface potential issues regarding some of the logic behind these regulations. Mapping the rules using data science techniques can help to consider them more holistically.

The unfamiliarity with the regulations often makes it hard for innovators to engage with them during the early research and development (R&D) stages. Health service providers are particularly well placed to comment on the R&D routes that medical devices take when they enter the clinical setting. It seems that a lot of new medical technology reaches the UK health service provider through non-commercial studies [10]. This is a thought-provoking finding, as a commercial company normally brings these medical technologies into the market. Delays in translation might occur if these non-commercial studies are not or less aware of the regulations. It should also be noted that only a very small number of these clinical studies seem to relate to software as a medical device (defined as a device that is entirely composed of software without any additional hardware). This poses a fascinating question in terms of how fast the field of software as a medical device is really growing. Looking at the number of devices that are registered in Australia, as software as a medical device, we found that there is indeed an upward trend [11]. However, these data from a publicly accessible database also made clear that software as a medical device only made up 1.6% of the total number of registered devices. It indicates that the majority of medical devices that are entering the market in this region are not software-based. These outcomes shed a more quantitative light on how fast stand-alone software with a medical purpose is moving into the market. Much of the research on medical Artificial Intelligence (AI) or Machine Learning (ML) might not yet have translated into a real market entry. This is likely because these methods are still relatively new from a regulatory standpoint, as well as the fact that software poses a different set of safety and performance problems compared to hardware. Nonetheless, it is important to look ahead and see how medical regulations might influence these new developments.

Another obvious topic that is gaining momentum is the environmental impact of healthcare innovation. The environmental impact of regulations on the product life cycle should be researched more thoroughly. Single-use devices and equipment are often selected to prevent pathogen transmission, but this tactic does come at an environmental cost [12]. More recently, the rising dependence on digital health records and information technology is starting to be mentioned in relationship to the environmental impact. Digital solutions might reduce landfill waste, but the energy requirements might create new challenges. These aspects will need to be considered along the more obvious waste management approaches of hardware.

In general, there is a strong need to take a more multi-disciplinary, holistic and data-driven approach in order to tackle the interconnected problems that emerge at the interface of regulations and medical technology. MedRegTech research allows for a critical appraisal of our current situation and could assist in the planning for the future. It can disrupt the regulatory landscape and help push the boundaries of our understanding forward to create better regulations for all.

Conflicts of Interest: The author declares no conflict of interest.

References

1. Ettlie, J.E.; Bridges, W.P. Environmental uncertainty and organizational technology policy. *IEEE Trans. Eng. Manag.* **1982**, *EM-29*, 2–10. [CrossRef]
2. Utterback, J.M.; Allen, T.J.; Hollomon, J.H.; Sirbu, M.A. The process of innovation in five industries in Europe and Japan. *IEEE Trans. Eng. Manag.* **1976**, *EM-23*, 3–9. [CrossRef]
3. Çanakoğlu, E.; Erzurumlu, S.S.; Erzurumlu, Y.O. How Data-Driven Entrepreneur Analyzes Imperfect Information for Business Opportunity Evaluation. *IEEE Trans. Eng. Manag.* **2018**, *65*, 604–617. [CrossRef]
4. Banisar, D.; Davies, S. Global Trends in Privacy Protection: An International Survey of Privacy, Data Protection, and Surveillance Laws and Developments. *John Marshall J. Comput. Inf. Law* **1999**, *18*, 1–112.
5. Gerstenfeld, A. Government regulation effects on the direction of innovation: A focus on performance standards. *IEEE Trans. Eng. Manag.* **1977**, *EM-24*, 82–86. [CrossRef]
6. Buckley, R.P.; Arner, D.W.; Zetzsche, D.A.; Weber, R.H. The road to RegTech: The (astonishing) example of the European Union. *J. Bank. Regul.* **2020**, *21*, 26–36. [CrossRef]
7. Arnould, A.; Hendricusdottir, R.; Bergmann, J. The Complexity of Medical Device Regulations Has Increased, as Assessed through Data-Driven Techniques. *Prosthesis* **2021**, *3*, 314–330. [CrossRef]
8. Mingay, H.R.F.; Hendricusdottir, R.; Ceross, A.; Bergmann, J.H.M. Using Rule-Based Decision Trees to Digitize Legislation. *Prosthesis* **2022**, *4*, 113–124. [CrossRef]
9. Bergmann, J.; Hendricusdottir, R.; Lee, R. Regulatory navigation: A digital tool to understand medical device classification pathways. *Compr. Biotechnol.* **2019**, *5*, 167–172.
10. Kwong, M.T.; Stell, D.; Akinluyi, E. Medical Device Regulation from a Health Service Provider's Perspective. *Prosthesis* **2021**, *3*, 261–266. [CrossRef]
11. Ceross, A.; Bergmann, J. Evaluating the Presence of Software-as-a-Medical-Device in the Australian Therapeutic Goods Register. *Prosthesis* **2021**, *3*, 221–228. [CrossRef]
12. Arun Kumar, P. Regulating Environmental Impact of Medical Devices in the United Kingdom—A Scoping Review. *Prosthesis* **2021**, *3*, 370–387. [CrossRef]

prosthesis

Article

The Complexity of Medical Device Regulations Has Increased, as Assessed through Data-Driven Techniques

Arthur Arnould, Rita Hendricusdottir and Jeroen Bergmann *

Natural Interaction Lab, Department of Engineering Science, University of Oxford, Oxford OX1 3PJ, UK; arthur.arnould@jesus.ox.ac.uk (A.A.); rita.hendricusdottir@eng.ox.ac.uk (R.H.)
* Correspondence: jeroen.bergmann@eng.ox.ac.uk; Tel.: +44-1865-273000

Abstract: Medical device regulations are dynamic, as they need to cover an ever changing landscape. In Europe this has led to a new set of regulations (both for Medical Devices and In Vitro Diagnostics), which replaced the old rules. This study is interested in how the complexity of these medical regulations changed over time and if additional time-based metrics can be associated with any of the complexity metrics. Complexity is defined in terms of readability of the text and it is computed using established linguistic measures, as well as Halstead complexity scores. It was shown that the regulatory complexity of new EU medical device regulations was higher than their predecessors, especially when Halstead complexity measures were considered. The complexity metrics obtained for the new regulations were subsequently associated with the time it took to consider these regulations. Only very weak Pearson's correlation coefficients were found between the complexity scores and the obtained response times for the new regulations. This could indicate that there are issues with how complexity is perceived by those that need to apply these regulations. Taking the complexity of regulations into account can greatly help with the development of more user friendly regulations. The results from the data-driven methods that are applied in this research indicate that governments could benefit from focusing on making regulations more accessible and utilitarian. This would improve the stakeholder adherence and facilitate effective implementation. This work also highlighted the need to develop more suitable methods to analyse regulatory text to further inform the wider research community.

Keywords: data science; regulations; law; medical devices; regulatory data science; natural language processing; linguistic analysis; optimisation

Citation: Arnould, A.; Hendricusdottir, R.; Bergmann, J. The Complexity of Medical Device Regulations Has Increased, as Assessed through Data-Driven Techniques. *Prosthesis* **2021**, 3, 314–330. https://doi.org/10.3390/prosthesis3040029

Academic Editor: Marco Cicciu

Received: 31 August 2021
Accepted: 24 September 2021
Published: 28 September 2021

1. Introduction

The medical device industry has been home to some of the most revolutionary innovations by mankind. The evolution in this field has been a cornerstone of the advances in global health. The term medical device itself encompasses a vast array of products that are used for diagnosing, treating and assisting patients. All aiming to improve the quality of life of patients. There is a great demand for new medical devices due to the aging population [1] and it is a growing global market, which was valued at $425.5 Billion in 2018 [2]. The United Kingdom alone was estimated in 2015 to have 4060 medical device manufacturers [3] and globally there are many small and medium enterprises operating in this domain. All these medical device companies have to engage with the medical device regulations if they want to bring their ideas to market.

Medical Device Regulations

Regulations are primarily designed to protect the patients, with regulatory bodies ensuring that medical devices are safe and that they perform as intended [4], but the regulations are also in place to protect the manufacturers themselves [5]. In some circumstances, patients may misuse devices or ignore instructions, injuring themselves in the process.

Compliance with the regulations acts as a safety net for the manufacturers in the case of a legal dispute. Manufacturers and designers can also use regulation to their advantage and innovate more efficiently, as they can follow clearer guidelines which in turn saves time and reduces some of the uncertainty regarding the process. Additionally, regulations enforce that devices offer a clinical benefit, which prevents markets from being flooded with devices that provide no gains for the user [6]. Finally, regulations are also intended to enhance post-market surveillance to ensure that devices have good longevity and prevents faulty devices from remaining in circulation [7].

While medical device regulations have been implemented for many years, much like other legislation, they do vary between different countries and continents. A major challenge faced by regulators is to ensure that the regulations remain relevant and fit for purpose. This is made more difficult by the sheer rate of innovation, as well as the ever changing needs within healthcare [8]. Consequently, regulations could become outdated and thus compromise patient safety. A prime example of this is the emergence of medical device software [9] and medical devices which have solely cosmetic purposes. These can range from coloured contact lenses to instruments used in cosmetic surgery for liposuction or tattoo removal. It was in fact a scandal related to leaky silicone breast implants from French company Poly Implant Prothèse that highlighted the short-comings in the Medical Device Directive which covered the regulations at that time [10]. Another well-documented example of where regulations did not adequately prevented harm was the discovery of blood poisoning in patients using metal-on-metal hip implants [11]. These kind of events led to an overhaul of the regulatory system in 2017 in Europe. Two new EU legislations were brought into force. Firstly the Medical Device Regulations (MDR) [12] replaced the Medical Devices Directive and Active Implantable Medical Devices Directive (MDD) [13,14]. Whilst, the In Vitro Diagnostic Regulations (IVDR) [15] replaced the EU [16,17]. For these new legislations, manufacturers had until May 2021 and May 2022 respectively to update their technical documents in order to conform with the new requirements. This clearly shows how manufacturers or suppliers of medical devices must continuously adhere to new regulations by ensuring that the correct device classification is achieved, general safety and performance requirements are met and conditions for clinical evidence are met. Manufacturers are responsible for regulatory compliance and that the specific requirements for importers, as well as distributors are respectively met where relevant [18].

2. Background

2.1. Complexity of Regulations

There are clear arguments for more stringent regulations in terms of patient safety, yet this form of legislation can only be effective if it is well understood and properly implemented. The MDR and IVDR are 175 and 157 page-long documents. They take the form of complex legal documents filled with lots of jargon. They can be difficult to navigate with the aim of locating specific information and subsequently the information itself can be hard to interpret once it has been found. Many medical devices, especially at the point of conceptualisation, are designed and manufactured by small or medium companies. The small nature of many companies means that their teams are often compact, with few employees. This means that they can struggle to secure the resources necessary to have sufficient legal competencies to navigate these regulations. Most of the innovators are academics or entrepreneurs with no or limited legal training or knowledge. Not understanding the regulations can cause misclassification and innovators might incorrectly interpret the regulations. This can have a compounding effect that leads to expensive redesign and/or retesting of certain devices. Not only does this hamper and discourage innovation, but companies could inadvertently develop faulty or inadequate devices. This raises concerns about patient safety, whilst simultaneously presenting risks to the manufacturers who have legal liabilities and also face significant losses of time and money. These outcomes are undesirable for all stakeholders and so it is vital to facilitate proper understanding of the regulations by medical device companies and innovators. The complexity of the legal text will have a direct effect on the ability of the reader to understand it.

2.2. *Complexity of Text*

The difficulty presented by a particular piece of text can be attributed to its linguistic complexity which is a measure of the extent to which the type of language used makes communication more or less complicated. Though the complexity of text is inherently subjective, for research purposes and large scale projects, it is important to utilise objective metrics to gain a better understanding of the problems innovators face.

Natural Language Processing (NLP) is a branch of computer science, combined with linguistics and artificial intelligence to interpret, analyse and process human text and speech across many languages. Recent progress in the management of unstructured data, the type generated from conversations or human written text, has equipped machines to understand language in a cognitive way which facilitates the identification of particular nuances and features of language [19]. The automatic manipulation of natural language by software is a field which has benefited from a data-driven approach.

The use of NLP methods enables the evaluation of text to return a quantitative score for its complexity and is particularly useful for treating vast quantities of data efficiently. These techniques range from numerical methods that focus on text length and sentence structure, to readability metrics which estimate the number of years of education required to easily read the text and finally, more evolved complexity methods that investigate the types of words and language used in a text to determine the overall complexity. Each of these methods are used in a range of fields, but not all are applicable to legal texts.

Intuitively, it can be assumed that the longer a piece of text is, the bigger of a challenge it will present and therefore it will require more time to read and process. This line of thinking generates numerical techniques which count the words, characters or syllables and only focus on the overall length as a predictor of expected time taken. However, the most common method for assessing text complexity originates from techniques focused around the expected level of education required to read the text. Traditionally, text complexity has been equated to its readability and this is what a series of methods from the 20th century aim to measure.

2.2.1. Common Readability Metrics

The Dale–Chall Readability formula was initially developed in 1948 under the name of "A Formula for Predicting Readability" [20], but has since been updated to reflect the changes undergone by language use [21]. The formula is based around a list of common words which are deemed "not difficult"; this list was initially 973 words long but has since been expanded to 3000 words. The formula combines the average length of sentences, in terms of the number of words, with the percentage of words which are not present on the list (these are known as difficult words). The method returns a grade estimate ranging from "Grade 4 and below" to "Grades 16 and above" which are equivalent to college graduates. The theory behind the use of familiar words as a metric, as opposed to letter or syllable count, is that tests have shown that readers typically find it easier to read, process and recall a passage if it is made up of familiar words [22,23]. However, this method has been criticised for failing to account for more complex structural relations within a text [24].

The Flesch-Kincaid Reading Ease formula follows a similar concept of determining the age of students who should be able to easily read a text [25]. This is achieved by attributing a score of 0–100 to the text; the higher the score, the easier the text. These scores are then associated with grades; scores above 90 indicate that the text is very easy and should be easily read by an average 5th grade student. The Flesch-Kincaid formula has become the chosen readability metric of many US Government Agencies such as the US Department of Defense. Also developed in 1948, the formula uses the ratios of words per sentence and syllables per word to calculate the score [26].

Another metric that assesses the U.S. grade level required to read sections of text is the Automated Readability Index (ARI) [27]. The index was designed in 1967 for real time readability on typewriters used by the military [28]. The origins of the formula dates back to the writing of manuals in the U.S. Navy, as the manuals used previously were written in a style which was above the reading capabilities of most of the staff. The ARI was validated as being more reliable and better suited to the technical nature of the text than other formulae such as Flesch-Kincaid [29]. Similarly, the ARI considers the words per sentence ratio, but it also includes the number of characters per word. This outputs a score that corresponds to the grade level required, but it exceeds the actual number of grades and reaches 14 (with this score corresponding to being above the level of a college student, for example a professor).

The Coleman–Liau index is the youngest of these formulae and was developed in 1975 to assess the readability of textbooks used in U.S. public schools [30]. Its creators deemed that counting syllables was too time-consuming and lacked accuracy. Therefore the Coleman–Liau Index uses the average number of letters per hundred words and also the number of sentences per hundred words [31]. This does have the drawback that it means that for shorter texts, these figures need to be extrapolated and thus may not be as representative. Once again, the numerical output is the estimated grade level required to read the text.

Whilst the Dale–Chal considers the familiarity of words, the other methods almost exclusively inspect the text as a collection of characters without considering the meaning of each word. Sentiment analysis is an area of machine learning which has seen exponentially more use in recent years. It aims to identify and extract subjective information from text which allows it to determine whether there is a positive, neutral or negative sentiment [32]. This process of perception is unconscious cognition within humans but until very recently was impossible for machines to achieve. Sentiment analysis techniques are incredibly powerful and are now being leveraged in fields such as market research, customer interactions and the analysis of social media activity [33]. Its entire premise is using computational linguistics to extract subjective meaning and information from text. In the regulatory context, all text is written in an objective manner to clearly outline regulations that must be adhered to and so sentiment analysis is redundant. Similarly, work done on phonetic analysis cannot be applied to this area either as it assesses speech rather than written text.

With this in mind, when exploring methods from other fields, it is important to consider the type of language that is used. Consequently, attention should be focused on similarly technical fields. The field of financial regulation was found to offer a wealth of previous work. This is largely due to an overhaul of financial regulation following the financial crisis of 2008. This increasingly stringent regulation has drawn attention, with a number of research papers investigating the change in complexity such as those by Gai et al. [34], Colliard & Georg [35] and Spatt [36]. Parallels can be drawn between this situation and the change in medical device regulation, therefore, the techniques should be transferable to a certain extent.

Historically, it was widely accepted in the field of financial research that the Gunning Fog Index was most suitable to measure the readability of documents and it was therefore almost universally used. First published in 1952 [37], the Gunning Fog Index is a metric that generates a grade level from 0 to 20 to indicate the level of education required to read the text, much like the other readability methods described earlier [38]. The formula combines the total number of words, number of sentences and also the number of complex words. Complex words are considered to be those consisting of three or more syllables, excluding common suffixes such as -es, -ed, or -ing. This list also excludes proper nouns, familiar jargon or compound words. A number of studies use the Gunning Fog Index, showing its popularity [39,40].

Despite being widely adopted, the method is not perfectly suited to for example financial text and this was highlighted by Loughran and Mcdonald [41]. At one point the Securities and Exchange Commission (SEC) considered using the Gunning Fog Index to gauge filings' compliance with the SEC's plain English initiatives, however, it was argued that it is ill-suited to analysing financial text, which inherently contains many longer words, despite these being well understood by analysts. Instead it was proposed that a focus on financial terminology and vocabulary that appears in a glossary and master dictionary is more pertinent to assess the readability. Though its efficacy may be questioned in a financial context, the Gunning Fog Index could be an interesting method to apply with the regulatory medical data field.

Whilst text containing intricate ways of describing concepts and elaborate language may once have been highly regarded and considered well written, there is now an ever growing desire for simplicity and effective communication in all fields. A prominent example of this is the Plain English Movement and other campaigns to limit the use of superfluous language and make technical text more accessible for everyone [42]. In some cases, these campaigns have published guidelines for individuals to refer to when drafting text to ensure that it is made as rudimentary as possible. Moreover, this drive for simple language extends further to regulatory agencies such as the SEC who provide very specific guidance in recommending that managers employ plain English attributes, by avoiding writing constructs like passive voice, weak or hidden verbs, superfluous words, legal and financial jargon, numerous defined terms, abstract words, unnecessary details, lengthy sentences, and unreadable design and layout in their financial disclosures [43]. The notion that the absence of such constructs makes text less complex (in a variety of fields that include medical, legal and military), is supported by many language experts [44]. It is with these considerations in mind that S.B. Bonsall IV et al. introduced a new readability metric by the name of the Bog Index [45]. The Bog Index aims to implement the concepts discussed above and one of its features is the way in which complexity is determined. The word complexity is derived from the principle of familiarity which is based on a proprietary list of over 200,000 words. This is in contrast to other techniques which assume that words are complex if they are multi-syllabic or contain many characters. Note that this is the primary criticism of the Gunning Fog Index from many language experts. The fundamentals of the Bog Index evaluate complexity using the trade off between Bog and Pep characteristics. Bog characteristics, as the name describes, bog the user down in unnecessary complexity such as jargon. Conversely, Pep identifies writing attributes that facilitate the understanding of texts by readers. The lower the Bog score, the easier the text is to read. 0–20 is considered excellent, most business and government writing scores 60–100 but some legal texts score over 1000 [46].

Colliard and Georg [35] also aimed to quantify the complexity of financial regulation by methods other than the mere length. In their work they attempted to achieve this by treating the regulation as an algorithm, using concepts from computer science literature to consider the rules for how an input leads to an output (the regulatory decision). The concept of operators and operands is the core feature of this analogy. This approach to complexity was pioneered by Maurice Howard Halstead [47]. The principle is that by segmenting a computer programme into its constituent parts, the relationships between these entities can be used to measure the algorithmic complexity. The two classes are known as the operators and operands and this logic is applied to financial regulation by [35]. Several techniques translate financial information such as balance sheets into pseudo-code to implement the algorithms developed. These methods will not be considered here due to the discrepancies in the format of the regulations. Instead, focus will be placed on methods designed for treating text. Words can be classified according to their function as either operands or operators, using the classification system proposed by Colliard and Georg. The focus is then to translate the algorithmic complexity from code to text-based analysis. The operators are words such as "and" or "excluding" which serve as logical connectors within an algorithm or, in this case, a sentence. "Operands" on the other hand are variables and parameters represented by values (e.g., "seven years" and "10 days"), concepts (e.g., "maturity" and "expiry") or entities (e.g., "manufacturers" and "council"). Words used for grammatical reasons (e.g., "by", "on", "the") can be ignored as they don't correspond to either operators or operands. Instead they are classed as function words which serve to ensure coherence of the text. The concept of unique operators and operands refers to words from those categories that have not been previously used in the text up to that point. A series of formulae relating the quantities of operators, operands, unique operators and unique operands were first developed by Halstead, but have since been tailored and added to by other papers including research by David Flater [48]. A combination of these equations can be used to assess the complexity of the regulatory text.

2.2.2. Complexity and Response Time

Time driven methods can provide a further insight into the complexity of the regulations in addition to metrics that are obtained directly from the available text. Text length is often positively associated with overall complexity and therefore response time. This theory also translates to question length, with longer questions requiring more cognitive resources and therefore being more likely to interfere with the mapping process. This suggests that question length is likely to be positively associated with the prevalence of both comprehension and mapping difficulties. However, if a question is long because the author has taken care to explain its intent fully, then comprehension will actually improve [49]. The medical device regulations can be posed as a set of questions [50], which allows for further exploration of this metric. One can then even incorporate a suggested minimum of 3 s that is considered needed to perform cognitive tasks and formulate responses to questions [51]. Applying this to the context of questions can provide a sense of expected response times once the reading time and answer selection time is accounted for.

Time-based data for medical regulations can be obtained from a rule-based classifier that is available online [52]. The questions have been ordered using a rule-based decision tree which leads the user through the nodes within the tree, ensuring that users only encounter questions that will aid the classification of their device, based on their previous responses. The technical text is made more accessible through the use of glossaries and examples to contextualise the information. The terms contained within the glossary are those which are defined individually in the definitions sections of the MDR and IVDR regulations. To accurately represent the classification guidelines, phrases of each question posed in the digital tool retain wording from the regulations, which provides a clear mapping back to the regulatory content published by the EU. Each user's interaction with the tool is recorded, which creates a unique database (that will be referred to as OGGD in this paper). Due to the importance of speed and simplicity in the classification process, the time taken to respond to each rule in the decision tree is used as a metric for the regulatory burden that each question places on the user.

2.2.3. Research Aims

The first aim of this paper is to explore how the complexity of the medical device regulations has changed when the MDD/IVDD was replaced with the new MDR/IVDR. The aforementioned complexity metrics can be applied to objectively assess this and determine to what extend regulations within the EU might have increased in complexity. Secondly, an association between the time it takes to consider parts of the regulation and the complexity of these parts will be investigated. This would provide an idea on how the complexity of regulations maps onto the user experience. It also provides an additional metric with time itself acting as an surrugate for complexity.

3. Materials and Methods

3.1. Linguistic Complexity of Regulatory Documentation

A descriptive analysis will be performed on the available textual data. This will include the total number of pages, letters, words and syllables. In addition, the average word length will be determined for each relevant document.

Subsequently, a set of readability metric scores will be computed consisting of the Dale–Chall Readability score, Automated Readability Index (ARI), Coleman Liau Index,

Gunning Fog, Flesch Kincade Grade and Bog Index. The formulae for these can be found in the Table 1. In the case of this paper, the output from each formula will be retained in its raw form for analysis as it is the variation of scores between questions that is of interest rather than the variation of estimates across individual techniques.

Table 1. Readability metrics. WC = Word Count, DWC = Difficult or Complex Word Count, SC = Sentence Count, SyC = Syllable Count, CC = Character Count.

Complexity Method	Equation	
Dale–Chall Readability Formula [21]	$0.1579 \times (\frac{DWC}{WC} \times 100) + 0.0496 \times (\frac{WC}{SC})$	(1)
Flesch Kincaid Grade Level [25]	$0.39 \times (\frac{WC}{SC}) + 11.8 \times (\frac{SyC}{WC}) - 15.59$	(2)
Automated Readability Index Formula [28]	$4.71 \times (\frac{CC}{WC}) + 0.5 \times (\frac{WC}{SC}) - 21.43$	(3)
Coleman Liau Index Formula (adapted from [31])	$5.89 \times (\frac{CC}{WC}) - 0.3 \times (\frac{SC}{WC}) - 15.8$	(4)
Gunning Fog Formula [31]	$0.4 \times [(\frac{WC}{SC} + 100 \times (\frac{DWC}{WC})]$	(5)

The Bog Index will also be determined, including contrasts Bog and Pep features which ultimately determine the Bog score for the text [45]. The StyleWriter software [46] is used to process the text in order to obtain Bog metrics.

$$Bog\ Index\ Formula = Sentence\ Bog + Word\ Bog - Pep \tag{6}$$

$$Sentence\ Bog = \frac{(\ average\ sentence\ length)^2}{long\ sentence\ limit} \tag{7}$$

$$Word\ Bog = \frac{(style\ problems + heavy\ words + abbreviations + specialist\ words) \times 250}{number\ of\ words} \tag{8}$$

$$Pep = \frac{(names + interest\ words + conversational) \times 25}{number\ of\ words} + sentence\ variety \tag{9}$$

For the Halstead-based methods, the first step was to classify words as either operators or operands. However, inferring the class of the elements of medical device regulations is novel and there is no preceding literature which can be used as a benchmark. Online scientific glossaries were used alongside the financial classification lists published by [35] and some case-by-case discretion to compile two distinct dictionary lists of words. Combined, these contained over 100,000 words, primarily medical in nature, which were utilised to assign classes. Words that were not contained within either the list of operators or that of operands were assigned to the "other words" category. These contained primarily function words that are included to make the text readable and coherent. The count of unique operators n_1, total operators N_1, unique operands n_2 and total operands N_2 formed the basis of the different approaches to quantifying complexity in the text (defined as "programme" for this approach). The metrics for complexity begin in a similar vein to the numerical techniques by associating psychological complexity to the length of the parts of the algorithm. Programme length, vocabulary size and programme volume, which actually measures the length of the binary encoding of the programme in software, all fall into this category. The remaining methods are in form of the ratio of total operators to total operand and the two programme level constructs, which take into consideration the most efficient expression of the programme by considering unique words. The equations for these methods are outlined below.

$$Programme\ Length,\ N = N_1 + N_2 \quad (10)$$

$$Vocabulary\ Size,\ n = n_1 + n_2 \quad (11)$$

$$Programme\ Level,\ L = \frac{(N_1 + N_2)}{(2 + n_2)} \quad (12)$$

$$Surrogate\ Programme\ Level,\ \hat{L} = \frac{(n_1 \times N_2)}{(2 \times n_2)} \quad (13)$$

$$Classification\ Ratio,\ C = \frac{N_1}{N_2} \quad (14)$$

$$Programme\ Volume,\ V = N \log_2 n \quad (15)$$

3.2. Data

Complexity of the text was determined for the European Union regulatory documentation that consisted of the new Medical Device Regulations (MDR) [12]/the In Vitro Diagnostic Regulations (IVDR) [15] and the older Medical Devices Directive (MDD) [13,14]/In Vitro Diagnostics Directive (IVDD) [16,17].

The time data was obtained from a online tool that mapped the MDR and IVDR on to a rule-based classifier [52]. Complexity of the text was determined for each node of the classifier. These nodes consisted of questions that cover specific parts of the regulations. The response time reflected the time it took to consider the relevant regulatory text and answer the question accordingly. This response data was generated using timestamps from users' interactions with the online regulatory text [52]. Upon submitting a response to a question on the corresponding web page, the user is then brought to a new web page corresponding to the next question to be answered according to the rule-based decision tree nodes. This submission generates an entry into a database which records the question number, the number of questions answered up to that point in the session, the answer submitted, the user's unique identification (ID) and the session ID. The time of answer submission is recorded as a timestamp and it is the difference between the timestamps for two successive questions that is used to calculate the time spent answering a particular question. In total the data set contains information for 903 unique user sessions, covering 112 questions. Data anonymization was performed, so there was no identifiable information about the users that generated the data.

3.3. Data Processing and Analysis

All readability metrics of the text were divided by taking the median of the full data set for a particular metric in order to aid the visual comparison. The same processing took place for both the data obtained from the legal documentation, as well as the online tool.

All outliers for the time data, in the form of response times less than 0.1 s in length and those greater than an hour, were removed. A total of 17,903 response data points were remaining. These were generated from 903 unique user session IDs, with a mean number of questions per session of 19.8. They covered responses to 112 nodes from the online tool. The time data had skewed distribution and thus the median was used for the analysis of the response times which were calculated for each node in the classifier. The readability metrics are computed for each of the 112 questions that are included in the online tool and represent the text in both the IVDR and MDR.

Pearson's Correlation Coefficients were computed between response times and the aforementioned complexity metrics. Data processing and analysis were done using Python (3.8.0, Python Software Foundation, Wilmington, DE, USA).

4. Results

4.1. Linguistic Complexity of Regulatory Documentation

The overall count of the MDR/IVDR compared to the MDD/IVDD has increased according to every metric, with some increase in the average word length as well (Figure 1).

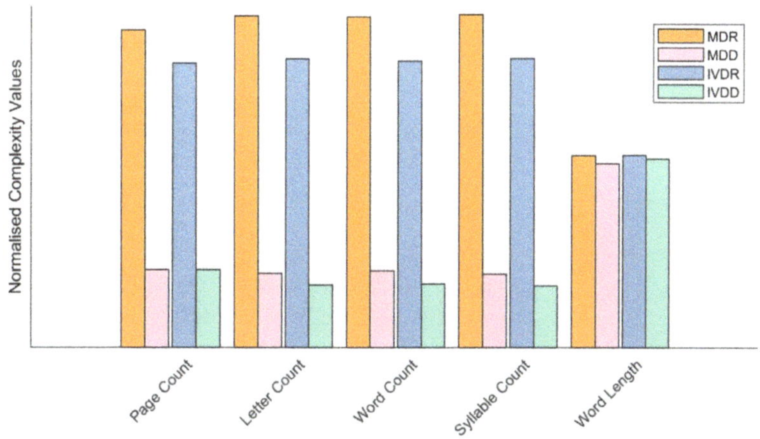

Figure 1. Descriptive complexity scores for the MDR, MDD, IVDR and IVDD. The word length represents the average word length.

The outcomes of the readability scores applied to the overall documents are shown in Figure 2.

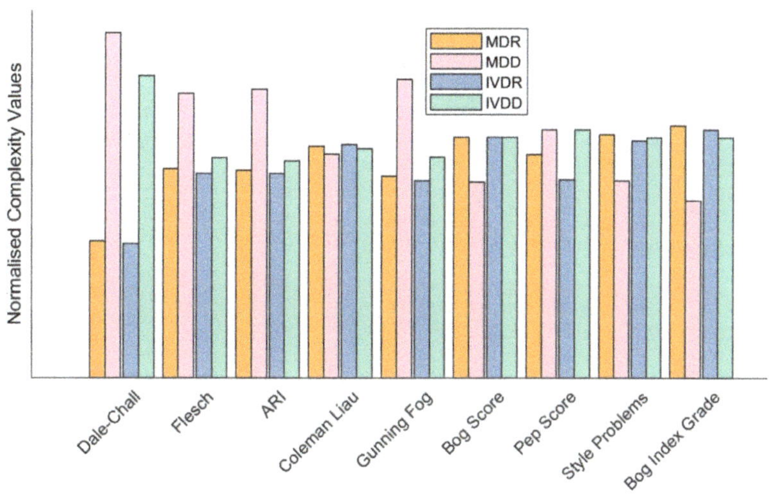

Figure 2. Readability scores for the MDR, MDD, IVDR and IVDD. Outcomes consist of the Dale–Chall Readability score, Automated Readability Index (ARI), Coleman Liau Index, Gunning Fog, Flesch Kincade Grade and Bog Index.

The Halstead-based complexity analysis is shown in Figure 3. All the metrics increased when new regulations (MDR/IVDR) were compared to the previous legislation.

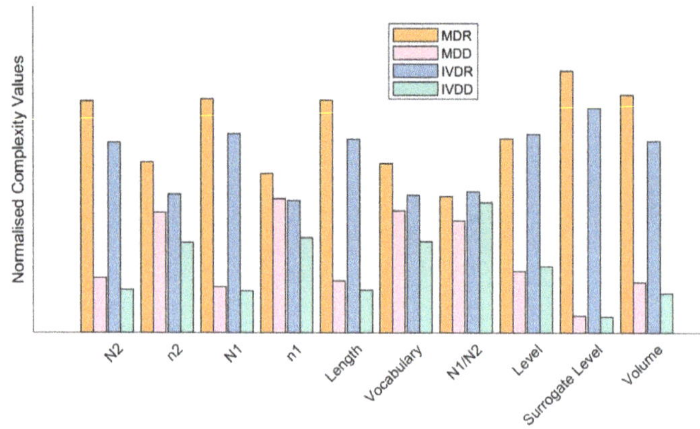

Figure 3. Halstead-based metric scores for the MDR, MDD, IVDR and IVDD. The following metrics are used: Total operators (N_1), unique operators count (n_1), total operands (N_2) and unique operands count (n_2). Other metrics consist of the length, vocabulary size, classification ratio (N_1 / N_2), level, surrogate level and volume of the "programme".

4.2. *Response Time and Linguistic Complexity*

Having removed the outliers, in the form of response times less than 0.1 s in length and those greater than an hour, there were 17,903 response times remaining. The majority (96.7%) of question/node response times were within a minute, with the median being 7 s for the overall dataset. Figure 4 shows a box and whisker plot for all of the nodes.

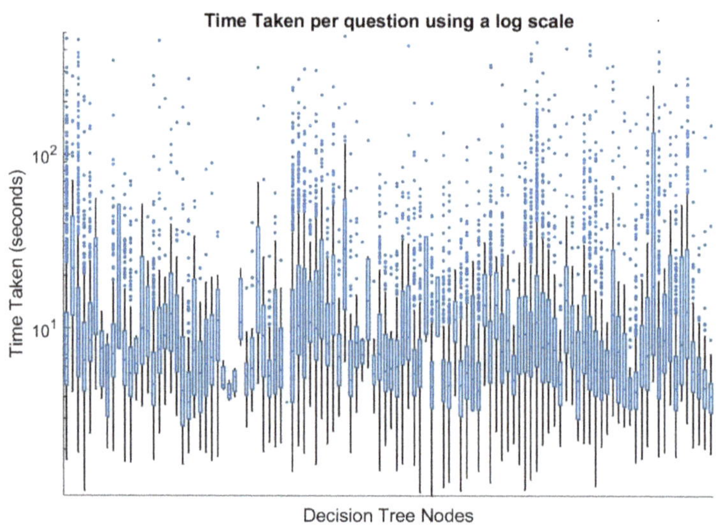

Figure 4. Box and whisker plot for the response times (vertical axis) of individual nodes (horizontal axis). Any data point which is greater than 1.5 times the interquartile range away from either the upper or lower quartile is denoted an outlier and marked by a circle. The data is displayed on a log axis for response time.

Figure 5 shows the median response time for each question plotted as a bar graph which is overlaid by the normalised values for three numerical complexity methods: Word length, letter count and word count. The variation in mean word length is smaller than the differences in overall question text length.

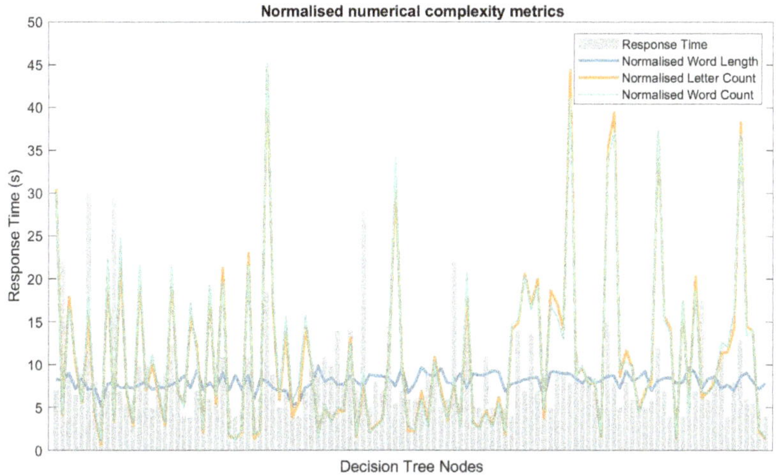

Figure 5. Normalised numerical complexity metrics and response times (vertical axis) of individual nodes (horizontal axis). The mean word length and median response times are shown. The central tendency metrics are selected based on the distribution of the data.

The readability scores for all 112 questions is shown in Figure 6. The histograms are overlaid on top of one another and there is a range of spread between the distribution of scores.

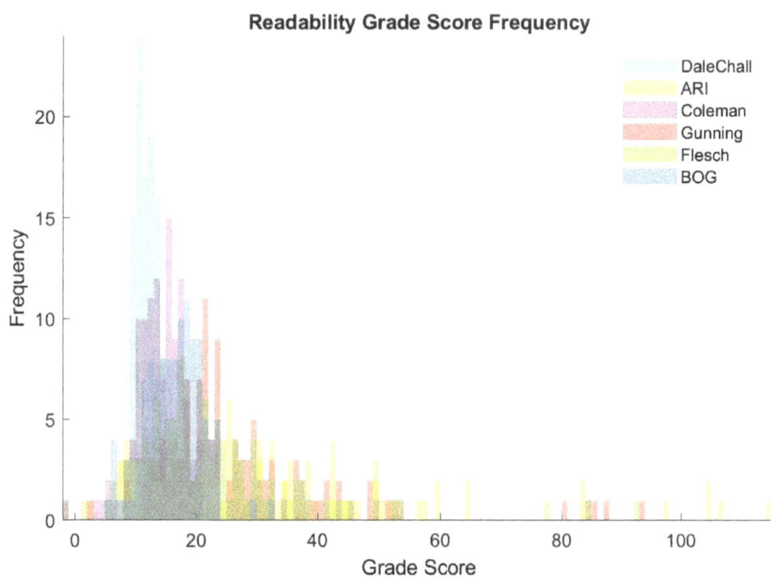

Figure 6. Frequency distributions for the readability metrics across all 112 nodes/questions of the digital tool.

The correlation coefficients for each of the readability methods are tabulated in Table 2. The Dale–Chall has the lowest absolute Pearson's correlation coefficient, whilst the strongest association is found between the word count and response time.

Table 2. Pearson's correlation coefficients for readability metrics and response times.

Complexity Metric	Pearson's Correlation Coefficient (p-Value)
Letter Count	0.204 (0.015)
Word Count	0.214 (0.011)
Syllable Count	0.202 (0.016)
Syllables per Word	−0.087 (0.180)
Letters per Word	−0.019 (0.421)
Dale–Chall	0.003 (0.488)
ARI	0.158 (0.048)
Coleman Liau	0.055 (0.283)
Gunning Fog	0.133 (0.081)
Flesch Grade	0.149 (0.059)
Bog Index	0.153 (0.054)

Four Halstead-based complexity metrics for each question are plotted with the response times overlaid upon it as a bar graph (see Figure 7).

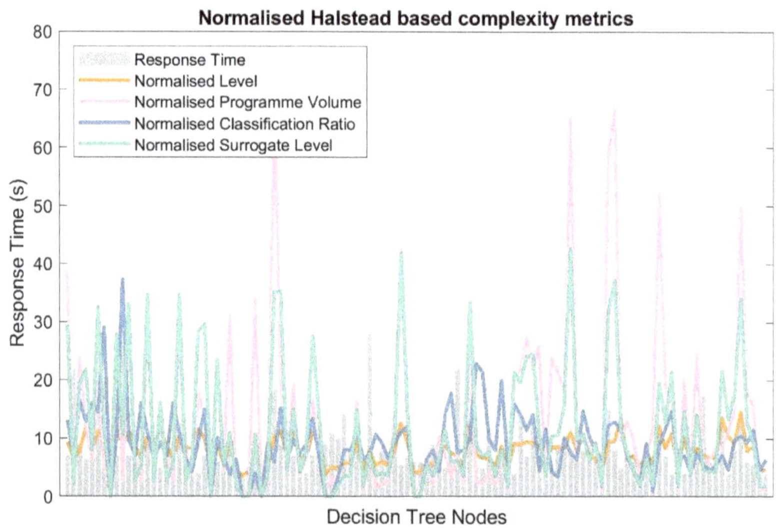

Figure 7. Normalised Halstead-based complexity metrics and response times (vertical axis) of individual nodes (horizontal axis).

The Pearson's correlation coefficients for Halstead-based complexity metrics are shown in Table 3. The lowest correlation coefficient was found for operand count and the highest was for the classification ratio. However, there was only a difference of 0.035 between these coefficients.

Table 3. Pearson's correlation coefficients (p-values) for Halstead-based complexity metrics and the response times.

Complexity Metrics	Pearson's Correlation Coefficient (p-Value)
Operator Count	0.200 (0.017)
Operand Count	0.193 (0.021)
Programme Length	0.200 (0.020)
Vocabulary Size	0.201 (0.020)
Classification Ratio	0.228 (0.008)
Level	0.209 (0.013)
Surrogate Level	0.214 (0.011)
Programme Volume	0.197 (0.019)

5. Discussion

In general, the majority of the complexity scores indicate that complexity increased with the implementation of the new legislation. The most important contributor to this seems to be the vast increase in length of the documents. This has undoubtedly made the regulations more difficult to negotiate for users.

5.1. Linguistic Complexity of Regulatory Documentation

The most complex words found in the regulations are in the form of technical medical references. While these words may be hard to understand, they are not particularly long. This is different to what is found for words in a non-scientific context where word length is indicative of complexity [53]. The findings of this study suggest that the conventional readability metrics may need to be further improved for use within a regulatory context and it echos the limitations summarised by Redish [54]. The techniques that were applied are actually used in many fields as the default measures of complexity. However, they are likely to fall short in similar ways as seen in this paper.

The Dale–Chall method, based on a set list of common words, did not yield a strong association with response time. This is understandable as the list of words was not designed with either scientific or technical objectives in mind and thus it would be constrained for this task. Perhaps a similar list, which had been tailored for for regulations could perform better. The Gunning Fog method also looks at a ratio of complex words and raises similar concerns as also seen in technical financial fields [45]. The Coleman Liau Index was designed to average metrics over 100 words, but several regulatory segments that need to be considered consisted of regulatory "questions" that were considerably shorter than this. Several approaches also heavily relied on the average word length, which might not capture complexity well in regulations. The ARI is the only method that was designed with technical content in mind, which should make it more suitable. However, it does not fully cover the regulatory context. Finally, there is the Bog Index which is far more recent and uses a more developed method to assess the complexity. This method gives importance to style and Pep, which is the way in which the language can serve the interest of the reader. Neither of these are considerations for writing regulation. Consequently, it was hard to differentiate between many of the regulatory text in this regard, which is reflected in the histograms of Bog Index scores. It should be noted that more sophisticated metrics need to be developed to capture the regulatory complexity more accurately. These metrics provide an initial assessment, but are limited in terms of measuring the nuances that are present within the regulatory text. Despite these limitations the metrics due seems to indicate a similar trend in terms of the complexity increase seen in the new regulations.

5.2. Response Time

The box plot with the individual nodes demonstrates how the spread of samples for each question varied (see Figure 4). These differences would be amplified without applying a log scale for the response time. Certain nodes have very tightly packed response times while others see a vast spread of times. This could be explained by the nature of each question. Some questions could be perceived as easy (or difficult) by most users and thus their response times will be very similar. However, other questions which are particularly technical or specific to a certain field may polarise the user cohort with some users having significant difficulties, which can explain the variation in ranges for response times. It is hard to account for this and perform robust corrections on the data. Yet, increasing the size of the data set could cover a more representative set of users.

The current data set consisted of over 17,000 unique response times. However, a larger data set can help further increase the external validity of the research findings. Furthermore, no in-depth information was gathered with regards to the response times that were generated, due to the fact that data was anonymised. Future studies can aim to collect information on the expertise of the user, device type considered and confidence of answering a particular question. This could create a better model, which can help to explain the response times that were observed.

5.3. Response Time and Linguistic Complexity

Only a (very) weak relationship was found between the response times and the complexity of the questions. This could be due to the lack of variation in the complexity measures. However, there are other factors that can also influence the strength of the association. Information was missing with regards to the specific intent of the device considered for the questions, the user's knowledge on regulations and a reliance on the outcome of the classifier. Therefore, the strength of these associations should not be considered to generalise easily. Response time data under more controlled settings can yield important additional information that can be used to build a more robust model to determine if there is not a stronger link between complexity and time. It should also be noted that the time spend on the questions might not be fully representative of the overall time spend on considering it. These are limitations that can be addressed in future studies.

5.4. Considerations

The techniques presented here offer a starting point to better understand the complexity of regulations. As shown there is no clear association between the time spent on a regulatory question and the associated complexity. However, it should be noted time only focuses the (instant) answer of the classification questions and therefore just captures the first steps in the regulatory process. More work can be done to objectively study the time consideration across the full application of the regulatory text. Using more data-driven methods can greatly increase our understanding of the regulations and allows to generate better questions that can help improve future regulations [9].

The complexity of a legal text can form a barrier to innovators. Understanding the complexity can therefore be essential in optimising the pathway for new devices. The increased complexity found in this study highlights the importance of improving education and guidance. Previous research already showed that more can be done to provide further support in terms of education [55] and this should be considered by stakeholders as complexity of regulations increases. Regulations that are currently being created can benefit from considering these aspects during the development process. These preliminary findings from this paper propose that the complexity of regulations should be reduced, with a focus on making them more accessible and utilitarian. This would improve the stakeholder adherence and facilitate effective implementation, which in the long term will improve patient welfare.

Author Contributions: Conceptualization, R.H. and J.B.; methodology, A.A. and J.B.; software, A.A. and J.B.; formal analysis, A.A.; investigation, A.A., R.H. and J.B.; resources, R.H. and J.B.; data curation, R.H. and J.B.; writing—original draft preparation, A.A. and J.B.; writing—review and editing, A.A., R.H. and J.B.; visualization, A.A.; project administration, R.H. and J.B.; funding acquisition, J.B. All authors have read and agreed to the published version of the manuscript.

Funding: This research was funded by the European Institute of Innovation and Technology (EIT) Health and supported by the National Institute for Health Research (NIHR) Oxford Biomedical Research Centre (BRC).

Institutional Review Board Statement: The study was approved by Central University Research Ethics Committee (CUREC): R63968/RE001.

Informed Consent Statement: Informed consent was obtained from all subjects involved in the study.

Data Availability Statement: All data on complexity were obtained from public documents and references are provided in the manuscript whenever they are discussed.

Conflicts of Interest: The authors declare no conflict of interest.

References

1. KPMG. The Changing Landscape of the Medical Devices Industry in the APAC Region. 2020. Available online: https://assets.kpmg/content/dam/kpmg/jp/pdf/2020/jp-medical-device-apac-en.pdf (accessed on 25 July 2021).
2. Insights, F.B. Medical Device Market Size, Share and Industry Analysis by Type. 2021. Available online: https://www.fortunebusinessinsights.com/industry-reports/medical-devices-market-100085 (accessed on 25 July 2021).
3. Office for Life Science. UK Medical Technology Sector, Bioscience and Health Technology Sector Statistics. 2019. Available online: https://assets.publishing.service.gov.uk/government/uploads/system/uploads/attachment_data/file/910441/Bioscience_and_Health_Technology_Statistics_2019_Infographic_-_Medical_Technology.pdf (accessed on 25 July 2021).
4. Em Agency. Medical Devices. Available online: https://www.ema.europa.eu/en/human-regulatory/overview/medical-devices (accessed on 25 July 2021).
5. World Health Organization. Medical Device Regulations: Global Overview And Guiding Principles. 2003. Available online: https://apps.who.int/iris/handle/10665/42744 (accessed on 25 July 2021).
6. Medical Device Coordination Group. Regulation (EU) 2017/745: Clinical Evidence Needed for Medical Devices Previously CE Marked under Directives 93/42/EEC or 90/385/EEC. A Guide for Manufacturers and Notified Bodies. Available online: https://ec.europa.eu/docsroom/documents/40904 (accessed on 25 July 2021).
7. World Health Organization. *Guidance for Post-Market Surveillance and Market Surveillance of Medical Devices, Including IVD*; World Health Organization: Geneva, Switzerland, 2020.
8. Soliman, E.; Mogefors, D.; Bergmann, J.H. Problem-driven innovation models for emerging technologies. *Health Technol.* **2020**, *10*, 1195–1206. [CrossRef]
9. Ceross, A.; Bergmann, J. Evaluating the Presence of Software-as-a-Medical-Device in the Australian Therapeutic Goods Register. *Prosthesis* **2021**, *3*, 221–228. [CrossRef]
10. Martindale, V.; Menache, A. The PIP scandal: An analysis of the process of quality control that failed to safeguard women from the health risks. *J. R. Soc. Med.* **2013**, *106*, 173–177.
11. News, D.I. The Metal-on-Metal Hip Implants Scandal. Available online: https://www.druginjurynews.com/news/metal-metal-hip-implants-scandal/ (accessed on 25 July 2021).
12. The European Parliament and the Council of the European Union. Regulation (EU) 2017/745 of the European Parliament and of the Council of 5 April 2017 on Medical Devices. 2017. Available online: https://eur-lex.europa.eu/legal-content/EN/TXT/PDF/?uri=CELEX:32017R0745 (accessed on 25 July 2021).
13. Medical Device Regulation Revision. Available online: https://www.bsigroup.com/en-GB/medical-devices/our-services/MDR-Revision/ (accessed on 25 July 2021).
14. The European Parliament and the Council of the European Union. Council Directive 93/42/EEC of 14 June 1993 Concerning Medical Devices. 1993. Available online: https://eur-lex.europa.eu/LexUriServ/LexUriServ.do?uri=CONSLEG:1993L0042:20071011:en:PDF (accessed on 25 July 2021).
15. The European Parliament and the Council of the European Union. Regulation (EU) 2017/746 of the European Parliament and of the Council of 5 April 2017 on In Vitro Diagnostic Medical Devices. 2017. Available online: https://eur-lex.europa.eu/eli/reg/2017/746/oj (accessed on 25 July 2021).
16. In Vitro Diagnostic Regulation Revision. Available online: https://www.bsigroup.com/en-GB/medical-devices/our-services/IVDR-Revision/ (accessed on 25 July 2021).

17. Directive 98/79/EC of the European Parliament and of the Council of 27 October 1998 on In Vitro Diagnostic Medical Devices. 1998. Available online: https://eur-lex.europa.eu/legal-content/EN/TXT/PDF/?uri=CELEX:31998L0079 (accessed on 25 July 2021).
18. Government, U. Guidance for Medical Devices: EU Regulations for MDR and IVDR (Northern Ireland). Available online: https://www.gov.uk/guidance/medical-devices-eu-regulations-for-mdr-and-ivdr (accessed on 25 July 2021).
19. Chowdhury, G.G. Natural language processing. *Annu. Rev. Inf. Sci. Technol.* **2003**, *37*, 51–89. [CrossRef]
20. Dale, E.; Chall, J.S. A formula for predicting readability: Instructions. *Educ. Res. Bull.* **1948**, *27*, 37–54.
21. Chall, J.S.; Dale, E. *Readability Revisited: The New Dale–Chall Readability Formula*; Brookline Books: Cambridge, MA, USA, 1995
22. Frisson, S.; Pickering, M.J. The processing of familiar and novel senses of a word: Why reading Dickens is easy but reading Needham can be hard. *Lang. Cogn. Process.* **2007**, *22*, 595–613. [CrossRef]
23. Zhang, J.; Liu, X.L.; So, M.; Reder, L.M. Familiarity acts as a reduction in objective complexity. *Mem. Cogn.* **2020**, *48*, 1376–1387. [CrossRef] [PubMed]
24. Benjamin, R.G. Reconstructing readability: Recent developments and recommendations in the analysis of text difficulty. *Educ. Psychol. Rev.* **2012**, *24*, 63–88. [CrossRef]
25. Schuyler, M.R. A readability formula program for use on microcomputers. *J. Read.* **1982**, *25*, 560–591.
26. Flesch, R. A new readability yardstick. *J. Appl. Psychol.* **1948**, *32*, 221. [CrossRef]
27. Senter, R.; Smith, E.A. *Automated Readability Index*; Technical Report; Cincinnati University: Cincinnati, OH, USA, 1967.
28. Thomas, G.; Hartley, R.D.; Kincaid, J.P. Test-retest and inter-analyst reliability of the automated readability index, Flesch reading ease score, and the fog count. *J. Read. Behav.* **1975**, *7*, 149–154. [CrossRef]
29. Smith, E.A.; Kincaid, J.P. Derivation and validation of the automated readability index for use with technical materials. *Hum. Factors* **1970**, *12*, 457–564. [CrossRef]
30. Coleman, M.; Liau, T.L. A computer readability formula designed for machine scoring. *J. Appl. Psychol.* **1975**, *60*, 283. [CrossRef]
31. Zhou, S.; Jeong, H.; Green, P.A. How Consistent Are the Best-Known Readability Equations in Estimating the Readability of Design Standards? *IEEE Trans. Prof. Commun.* **2017**, *60*, 97–111. [CrossRef]
32. Prabowo, R.; Thelwall, M. Sentiment analysis: A combined approach. *J. Inf.* **2009**, *3*, 143–157. [CrossRef]
33. Gohil, S.; Vuik, S.; Darzi, A. Sentiment analysis of health care tweets: review of the methods used. *JMIR Public Health Surveill.* **2018**, *4*, e5789. [CrossRef]
34. Gai, P.; Kemp, M.H.; Sánchez Serrano, A.; Schnabel, I. Regulatory Complexity and the Quest for Robust Regulation. Number 8. Reports of the Advisory Scientific Committee. 2019. Available online: https://ideas.repec.org/p/srk/srkasc/20198.html (accessed on 25 July 2021).
35. Colliard, J.E.; Georg, C.P. Measuring Regulatory Complexity. 2020. Available online: https://www.institutlouisbachelier.org/wp-content/uploads/2019/11/papier_jean-edouard-colliard.pdf (accessed on 25 July 2021).
36. Spatt, C.S. Complexity of regulation. *Harv. Bus. L. Rev. Online* **2012**, *3*, 1.
37. Gunning, R. *The Technique of Clear Writing*; McGraw-Hill: New York, NY, USA, 1952.
38. Bothun, L.S.; Feeder, S.E.; Poland, G.A. Readability of Participant Informed Consent Forms and Informational Documents From Phase III COVID-19 Vaccine Clinical Trials in the United States. In *Mayo Clinic Proceedings*; Elsevier: Amsterdam, The Netherlands, 2021. [CrossRef]
39. Li, F. Annual report readability, current earnings, and earnings persistence. *J. Account. Econ.* **2008**, *45*, 221–247. [CrossRef]
40. Miller, B.P. The effects of reporting complexity on small and large investor trading. *Account. Rev.* **2010**, *85*, 2107–2143. [CrossRef]
41. Loughran, T.; McDonald, B. Measuring readability in financial disclosures. *J. Financ.* **2014**, *69*, 1643–1671. [CrossRef]
42. Felsenfeld, C. The plain English movement. *Can. Bus. LJ* **1981**, *6*, 408.
43. Securities and Exchange Commission. Plain English Disclosure. 1998. Available online: https://www.federalregister.gov/documents/1998/02/06/98-2889/plain-english-disclosure (accessed on 25 July 2021).
44. DuBay, W.H. The Principles of Readability. Impact Information. 2004. Available online: http://impact-information.com/impactinfo/readability02.pdf (accessed on 25 July 2021).
45. Bonsall IV, S.B.; Leone, A.J.; Miller, B.P.; Rennekamp, K. A plain English measure of financial reporting readability. *J. Account. Econ.* **2017**, *63*, 329–357. [CrossRef]
46. Nirmaldasan. StyleWriter's Bog Index. Available online: https://strainindex.wordpress.com/2010/01/19/stylewriters-bog-index/m (accessed on 25 July 2021).
47. Halstead, M.H. *Elements of Software Science (Operating and Programming Systems Series)*; Elsevier Science Inc.: Amsterdam, The Netherlands, 1977.
48. Flater, D.W. *'Software Science' Revisited: Rationalizing Halstead's System Using Dimensionless Units*; National Institute of Standards and Technology: Gaithersburg, MD, USA, 2018.
49. Holbrook, A.; Cho, Y.I.; Johnson, T. The impact of question and respondent characteristics on comprehension and mapping difficulties. *Int. J. Public Opin. Q.* **2006**, *70*, 565–595. [CrossRef]
50. Bergmann, J.H.; Hendricusdottir, R.; Lee, R. Regulatory Navigation: A Digital Tool to Understand Medical Device Classification Pathways. In *Comprehensive Biotechnology*, 3rd ed.; Moo-Young, M., Ed.; Pergamon: Oxford, UK, 2019; pp. 167–172. [CrossRef]
51. Stahl, R.J. Using "Think-Time" and "Wait-Time" Skillfully in the Classroom. ERIC Digests. 1994. Available online: https://eric.ed.gov/?id=ED370885 (accessed on 25 July 2021).

52. Natural Interaction Lab, University of Oxford. Oxford Global Guidance. 2021. Available online: https://www.oxfordglobalguidance.org (accessed on 25 July 2021).
53. Lewis, M.L.; Frank, M.C. The length of words reflects their conceptual complexity. *Cognition* **2016**, *153*, 182–195. [CrossRef] [PubMed]
54. Redish, J. Readability formulas have even more limitations than Klare discusses. *ACM J. Comput. Doc. (JCD)* **2000**, *24*, 132–137. [CrossRef]
55. Hendricusdottir, R.; Hussain, A.; Milnthorpe, W.; Bergmann, J.H. Lack of Support in Medical Device Regulation within Academia. *Prosthesis* **2021**, *3*, 1–8. [CrossRef]

prosthesis

MDPI

Article

Using Rule-Based Decision Trees to Digitize Legislation

Henry R. F. Mingay, Rita Hendricusdottir, Aaron Ceross and Jeroen H. M. Bergmann *

Department of Engineering Science, University of Oxford, Parks Road, Oxford OX1 3PJ, UK;
henry.mingay@eng.ox.ac.uk (H.R.F.M.); rita.hendricusdottir@eng.ox.ac.uk (R.H.);
aaron.ceross@univ.ox.ac.uk (A.C.)
* Correspondence: jeroen.bergmann@eng.ox.ac.uk

Abstract: This article introduces a novel approach to digitize legislation using rule based-decision trees (RBDTs). As regulation is one of the major barriers to innovation, novel methods for helping stakeholders better understand, and conform to, legislation are becoming increasingly important. Newly introduced medical device regulation has resulted in an increased complexity of regulatory strategy for manufacturers, and the pressure on notified body resources to support this process is making this an increasing concern in industry. This paper explores a real-world classification problem that arises for medical device manufacturers when they want to be certified according to the In Vitro Diagnostic Regulation (IVDR). A modification to an existing RBDT algorithm is introduced (RBDT-1C) and a case study demonstrates how this method can be applied. The RBDT-1C algorithm is used to design a decision tree to classify IVD devices according to their risk-based classes: Class A, Class B, Class C and Class D. The applied RBDT-1C algorithm demonstrated accurate classification in-line with published ground-truth data. This approach should enable users to better understand the legislation, has informed policy makers about potential areas for future guidance, and allowed for the identification of errors in the regulations that have already been recognized and amended by the European Commission.

Keywords: classification; healthcare; innovation; regulation; medical devices; decision tree complexity

Citation: Mingay, H.R.F.; Hendricusdottir, R.; Ceross, A.; Bergmann, J.H.M. Using Rule-Based Decision Trees to Digitize Legislation. *Prosthesis* **2022**, *4*, 113–124. https://doi.org/10.3390/prosthesis4010012

Academic Editor: Marco Cicciu

Received: 8 January 2022
Accepted: 1 March 2022
Published: 10 March 2022

1. Introduction

Regulations contain rules setup by (governmental) authorities to control specific aspects of certain industries, which often influences the way companies operate. These rules affect how industries are managed, and the importance of regulations is such that many companies create specific divisions focused solely on the regulatory strategy [1]. Regulatory frameworks encourage consumers to adopt innovations by ensuring that their safety and effectiveness has been evaluated. Regulations promote better utilization of technologies and encourage the identification of novel technologies within industries [2]. When used effectively, regulation drives the direction of innovation [3], and can stimulate innovative ecosystems [4]. However, regulations also create barriers that can hold up the innovative process [5]. For innovative firms, they are one of the most significant barriers of perceived environmental uncertainty [6]. Organizational characteristics can have a significant influence on the effect of the regulation with regards to innovation [7]. SMEs need more information to better identify the relevant regulation and understand the requirements for conformity based on these regulations [8], which are the first steps in the regulatory navigation pathway. A better understanding of legislation can help to alleviate the barrier to innovation that regulation presents. One useful approach to provide a better understanding of complex legislation is by applying engineering methods to represent legislations in a more quantitative manner. Digitizing legislation is proposed as a method to present the rules within legislation in a simpler and logic-based format for the stakeholders that need to conform to the regulation. There are two types of classification problems resulting from legislation: identification of relevant regulation, and classification of certain use cases for conformity requirements within the legislation. While all products and services need to consider relevant legislation, a smaller subset of legislations contain risk-based classification problems within the legislative document. Examples of this include the classification of data types within data protection regulation (e.g., the General Data Protection Regulation) and the classification of devices within EU medical device regulation (e.g., the Medical Device Regulation and In Vitro Diagnostic Regulation). Legislations containing risk-based classes have different conformity requirements depending on the risk level of the use case in question. This article presents a novel method for using rule-based decision trees to digitize classification problems within legislation, using the In Vitro Diagnostic Regulation (IVDR) as a case study example. Rule-based decision trees were selected as classification is based on the rules within the legislation. It is not dependent on a specific training dataset. This use case will focus on the classification problem within the IVDR legislation, although it is also applicable for determining whether the devices are governed by the IVDR. These classifications problems appear in complex legislation with multiple conformity requirements. Stakeholders require greater understanding of these types of legislation as the conformity requirements are dependent on the risk level of their product or service.

2. Digitizing Legislation

Early attempts to create digital tools for classification problems within legislation were constrained by challenges arising from differences between legal and technical semantics. Where engineering lexicon is based on precision [9], legal language is characterized by its construction to remain as generalized as possible to the class of problems that the legislator is addressing [10]. Legal requirements present multiple problems for managing compliance including ambiguities, cross-referencing, acronyms, domain-specific definitions, and frequent amendments due to revisions of legislation and case law [11]. One of the first attempts to digitize legislation was undertaken by Sergot et al. (1986) [12], when the British Nationality Act was converted into a logic program. They concluded that representation of rule-based legislation using logic programming is optimal, as it is simple for both naive users and experts to understand. This method allowed users to infer meaning from the legislative rules and made modifications of the system straightforward. Translation of laws into a logic program was shown to provide a better understanding of the rules within the legislation and identified specific interpretation issues within legislation [13]. The use of logic programming for representation of legislation was met with a scathing rebuttal by Leith (1986) [14]. The professor of law argued that 'legislation could not be formalized in a truly logical format'. The use of a logic program was criticized for embedding assumptions into the logic program arising from vague concepts. Challenges regarding the interpretation and ambiguity within the legislation were highlighted. Other attempts to digitize legislation consisted of a proof of concept for modelling the Italian data protection legislation created by Massaccia et al. (2005) [15], and an evaluation tool for the compliance of websites to Canada's Personal Information Protection and Electronic Documents Act (PIPEDA) [16].

This article builds on the concept of using decision trees to better understand medical device regulation proposed by Bergmann et al. [17]. It will focus on the use of decision trees for representation of legislation, as they are suited to deal with issues arising from the differences between legal and technical semantics. Where engineers previously embedded assumptions into logic programs, decision trees allow users to directly evaluate the rules within the legislation at the level of attributes along the decision pathway.

3. Decision Trees for Digitizing Legislation

Bergmann et al. [17] proposed a data-driven methodology using the Iterative Dichotomiser 3 (ID3) decision tree learning methodology, introduced by Quinlan (1985) [18], to classify medical devices according to the rules set out in the Medical Device Directive (MDD). The alternative method to building a decision tree using a data-driven approach is the use of a rule-based approach. Knowledge is stored in the form of a set of rules, which are converted into a decision tree when a decision-making process is required [19]. Rule-based decision trees can be built from either static or dynamically changing decision rules [20]. If classification rules are stored as a set of declarative statements, then there are no constraints on the order in which the attributes can be evaluated, meaning that it is much easier to update the decision tree when new information and data are available. This is preferable for decision trees based on legislation, as there are often amendments added, revisions made in the legal text, and real-world examples published to further clarify ambiguities and issues with interpretation.

Decision trees also have advantages over 'black-box' classification models, including neural networks, as the logical rules in a decision tree are much easier to interpret than the complex relationship between hidden nodes in a neural network for the decision makers [21]. This allows stakeholders to obtain a one-to-one mapping between their product or service and the classification rules in the legislation, often required for conformity assessment. In addition to the ease of interpretability, a decision tree built from a pre-defined set of decision rules allows the classification of examples where there is no available data for training.

An important advantage that rule-based decision trees have over both data-driven decision trees and neural network approaches is the accuracy of the classifications from the legislative text. Classification is directly based on the rules that are set out in the legislation, and the accuracy of the classification is not dependent on training the decision tree using available data.

Overview of Rule Based Decision Tree Algorithms

The earliest paper identified that introduces the concept of a rule-based decision tree was Imam et al. [19]. The paper introduces a method called the AQDT-1 (AQ-derived Decision Tree-1), which builds a decision tree from decision rules generated using AQ-type inductive learning. The algorithm optimizes the order in which the rules are assessed in the decision tree structure. It is important to note that while this paper uses a set of rules that has been generated using inductive learning, the algorithm can be used to build a decision tree from any set of rules, including those in legislation. Rules at each node of the decision tree will subdivide the examples into smaller groups depending on whether the attribute at the node applies or not. The attributes are derived from the rules and placed at respective nodes along the decision tree. An attribute selection criterion is used to analyze the relationship between the attributes and the classification rules, and the order, in which they are evaluated, is optimized based on multiple selection criteria. This method for optimizing the ordering of the attributes in the decision tree is determined by an attribute utility ranking comprising of three criteria: disjointness, dominance, and extent. These, respectively, measure the effectiveness an attribute has in determining the final decision class for an example, the frequency in which this attribute is present in the rules and the number of values an attribute can take within the rules. The algorithm aims to maximize the disjointness and dominance of an attribute while minimizing its extent. Similar to greedy splitting decision trees, this method creates a large number of leaf nodes in the decision tree and decision rules are often pruned depending on the rule strength.

The AQDT-1 tree, whilst being novel in terms of the idea that it proposed, was far from optimal when evaluating the complexity of the decision trees created. This led to a publication shortly after by Michalski et al. [22], which outlined a refined method for this idea: the AQDT-2 algorithm. This method was shown to outperform the AQDT-1 in terms of both accuracy and decision tree complexity—determined by the number of decision tree leaf nodes. The notion of the cost of evaluating against certain attributes was introduced in this paper. It was deduced that the "inexpensive" attributes should be evaluated first, and thus should be assigned close to the root node. The "expensive" attributes should only be evaluated when necessary, therefore placed further away from the root node. This cost is quantified using an importance score. While this optimization was shown to improve the predictive accuracy and reduce the complexity of the decision tree, it was accompanied by the drawback that to calculate the cost of an attribute, a training data set is required.

The next rule-based approach was created when Abdelhalim et al. (2014) [20] introduced the concept of the Rule Based Decision Tree (RBDT-1) algorithm, which does not depend on a training dataset for optimization. This paper highlighted the fact that rule-based approaches are often used in situations with limited data, and commented that the previously developed AQDT-2 required a training set for optimization. The RBDT-1 algorithm depends on only the rules themselves and uses three parameters for optimization question sequencing: attribute effectiveness, attribute autonomy and minimum value distribution. Attribute effectiveness (AE) is the first of these to be considered. It depends on the influence that the attribute has in determining the class of an example. If an attribute does not contribute to the decision process of assigning a class to an example, then it has a lower AE score than one that does. The attribute with the highest AE score is chosen for the root node and then subsequent attributes are selected depending on the next highest AE score until a leaf node is reached. The attribute autonomy (AA) is considered when two attributes have the same AE score. This criterion selects attributes that create fewer subsequent nodes before the leaf node in an attempt to reduce the depth of the decision. The minimum value distribution (MVD) is only taken into account when two attributes have both the same AE and AA score. It also aims to minimize the complexity of the decision tree by favoring attributes that have fewer values in a given rule. An attribute with fewer values will in turn generate fewer branches, which will result in a smaller and less complex decision tree. Table 1 introduces an example problem, which is a modified version of a decision tree classification problem outlined in the literature [20]. Information is given in Table 1 for a range of different medical device companies, including the expertise level of the staff, size of the company (small <50 employees; medium 50–250 employees; large >250 employees) and investment attracted. The final column shows what level of (medical) risk the product developed by that specific company carries. More stringent regulations govern higher risk devices, which translates into additional costs and time for businesses.

Table 1. Device risk dataset used to compare the different decision tree algorithms. The "Expertise" column captures the human capital within a company with regards to medical devices. The "Size" indicates the number of employees within a company. The "Investment" shows how much investment that company had attracted, whilst the "Device risk" shows the risk of the medical devices they develop.

Company	Expertise	Size	Investment	Device Risk
1	No	Small	High	None
2	Yes	Small	High	Low
3	No	Large	High	None
4	No	Medium	High	None
5	Yes	Medium	Low	Medium
6	No	Medium	Low	None
7	Yes	Large	Low	None
8	Yes	Large	High	High
9	No	Large	High	None
10	Yes	Small	High	Low

The RBDT-1 is compared to previous rule-based and data-driven decision trees across a variety of decision tree classification problems, and is shown to outperform the AQDT-1, AQDT-2 as well as the ID3 (entropy driven) algorithms in terms of complexity, as shown in Figures 1 and 2. This is exhibited by the smaller number of attribute nodes and leaf nodes generated by the RBDT-1 decision tree. The RBDT-1 algorithm generated three attribute nodes and five leaf nodes, whilst the other algorithms ended up with five attribute nodes and seven leaf nodes.

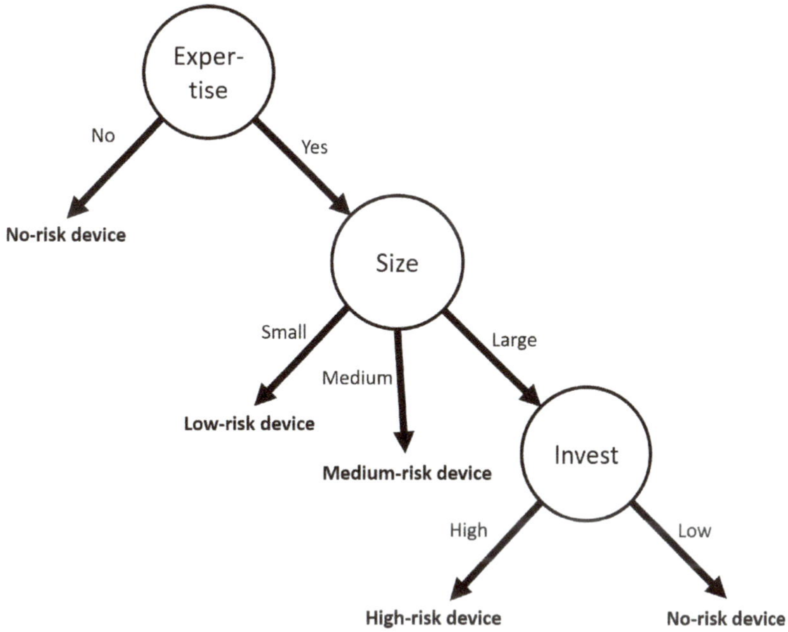

Figure 1. Decision tree generated for the device risk problem presented in Table 1 using the RBDT-1 algorithm [20]. It consists of three attribute nodes and five leaf nodes.

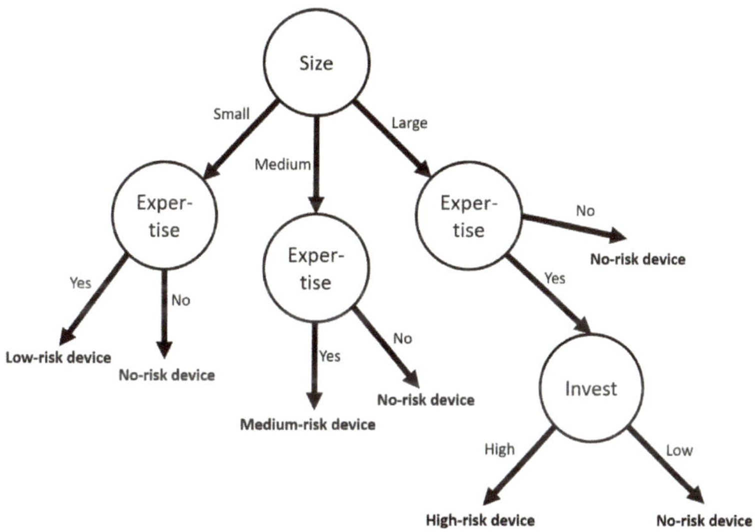

Figure 2. Decision tree generated for the device risk problem presented in Table 1 for the AQDT-1, AQDT-2 and ID3 algorithms [20]. It consists of five attribute nodes and seven leaf nodes.

While there is some literature introducing design of rule-based decision trees, there is a gap in the literature regarding the optimization of rule-based decision trees with additional case specific constraints. Currently, rule-based decision tree algorithms require modification to incorporate any additional constraints arising from the legislation. A rule-based decision tree that allows for constraints to be set (RBDT-1C) is introduced in this paper and the In Vitro Diagnostic Regulation (IVDR) is used as a case study example.

4. IVDR Legislation as a Case Study Example of RBDT-1C

Previous European medical device regulation promoted innovation by favoring faster regulatory approval and patient-access to novel technologies, which resulted in a series of patient outcome scandals—most notably regarding the Poly Implant Prosthesis (PIP). This led to a major overhaul of European medical device regulation in 2017, where the new regulation aims to increase the safety and effectiveness of medical devices by placing more emphasis on clinical evaluation and post market surveillance [23]. The new regulation has reacted to changes in the industry, including a rise in the use of medical device software [24]. However, Arnould et al. [5] have demonstrated that this has led to an increased complexity of Medical Device Regulations (both the MDR and IVDR). There is currently a shortage of resources within organizations responsible for certifying European medical devices due to the introduction of the new legislation [25], political factors [26], and other factors. The case study introduced in this paper is the use of a digital decision tree for classification of medical devices according to the new IVDR legislation. This can help to alleviate the strain for the regulatory stakeholders by providing resources to help medical device manufacturers classify their devices according to the new regulation. The proposed RBDT-1C algorithm was used to design a decision tree for the classification of in vitro diagnostic (IVD) devices into the risk-based classes outlined in the IVDR. The risk-based classes in the IVDR combine both patient and population risk as follows: Class A, low patient and population risk; Class B, moderate to low patient risk and low population risk; Class C, high patient risk and low population risk; and Class D, high patient and population risk. Population risk is important for IVD devices as a false negative (meaning the patient is positive) could endanger others if the disease is transmissible and life-threatening.

Decision tree attributes were created from the classification and implementing rules in Annex VIII of the IVDR [27]. Rules with multiple classification criteria were split into multiple attributes, as prior research has shown that complex questions lead to greater complexity of comprehension [28] and interfere with the mapping process [29], where the mapping process relates to the cognitive process, by which a user simultaneously recalls the question and answers based on the example they are testing. This additional complexity is likely due to the increased demand on the working memory of the user [30]. All complex and lengthy rules can be split into multiple attributes. Where rules contained many sub-rules, natural language processing (NLP) techniques were used to assess the semantic similarity of the sub-rules to cluster sub-rules into semantically meaningful groups. While long and complex questions are undesirable for a decision tree, excessive splitting of the sub-rules would lead to a deeper and more complex decision tree. Semantically similar sub-rules do not overcomplicate attributes, meaning multiple sub-rules can be asked at a single node of the decision tree if they relate to similar devices. The semantic properties of the sub-rules were assessed using word embedding methods, which convert words into vector-representations, capturing and displaying their semantic properties in the vector space. The word embeddings for each sub-rule were determined using the pre-trained BioBERT sentence embedding algorithm [31]. BioBERT's pretrained and fine-tuned deep neural network allows it to create vector representations where both the complex characteristics of words such as syntax and semantics, as well as the linguistic context of the word to model polysemy, are considered and reflected in the sentence embedding. BioBERT is also pre-trained on all PubMed data, meaning it can capture the semantic properties of biomedical specific terminology.

4.1. Building the RBDT-1C

After the attributes were created from the classification and implementing rules in the IVDR, the decision tree was built using a modified version of the RBDT-1 algorithm proposed by Abdelhalim et al. (2014) [20]. The use of rule-based decision trees for the representation of legislation is possible when using the RBDT-1 algorithm, as no data is required to train the decision tree. Additional constraints proposed by legislation can be designed into the decision tree by introducing supplementary criteria to the RBDT-1 algorithm, creating the RBDT-1C methodology—detailed in Supplementary Materials.

For the IVDR, the classification of a device is determined by the rule with the highest risk-based classification. If multiple rules applied to one device, then the rule with the highest risk-based classification determines the class of the device (IVDR, Annex VII, Implementing rule 1.9), where Class D is the highest risk-based classification, and Class A is the lowest. The risk-based classification is based on the personal risk and public health risk associated with an IVD device. The conformity constraint for the classification of devices is that all classification and implementing rules had to be considered when classifying a device (IVDR, Annex VIII, Implementing rule 1.7). This means that even if a higher risk-based rule applies, the lower risk-based rules must be assessed despite not contributing towards the final classification due to the "hierarchy" of classification. The conformity constraint was adapted into the tree using a novel attribute hierarchy criterion (AH). This is defined for each attribute as the number of subsequent attributes that can be excluded owing to the attribute taking a specific value. This is calculated multiple times for each attribute, once for each of the potential values that the attribute can take. These exclusions were then designed into the final decision tree structure, where the ordering of the IVDR attributes was determined by the three criteria within the original RBDT-1 algorithm: attribute effectiveness, attribute autonomy and minimum value distribution.

If devices are not independent medical devices (e.g., calibrators) they are classified according to the risk-level of the 'Parent Device'. The 'Parent Device' is the device that they are used with. The dotted line in Figure 3 represents the choice that the medical device manufacturer is given to proceed to classify the 'Parent Device' when a device does not have an independent classification. Additionally, Implementing Rule 1.4 (Table 2) refers to medical device software, which is only classified in its own right if this is independent of any other medical devices. If the software drives or influences the use of another medical device, then it is classified according to the 'Parent Device' that it is used with.

Table 2. IVDR Classification Rules dataset for the IVDR decision tree.

Rule	IVDR Classification Rule	Independent Medical Device	Personal Risk	Public Health Risk	Device Risk
7.1.1	Rule 7	Yes	Moderate/Low	Low	Class B
7.1.2	Implementing Rules 1.6	No	-	-	Parent device classification
7.2	Implementing Rules 1.5	No	-	-	Parent device classification
7.3.1	Implementing Rules 1.4	No	-	-	Parent device classification
7.3.2	Implementing Rules 1.4	Yes	-	-	Device classified in its own right
1.1	Rule 1	Yes	High	High	Class D
1.2	Rule 1	Yes	High	High	Class D
1.3	Rule 1	Yes	High	High	Class D
2.1	Rule 2	Yes	High	Moderate/Low	Class C
2.2	Rule 2	Yes	High	High	Class D
3.1	Rule 3	Yes	High	Moderate/Low	Class C
3.2	Rule 3	Yes	High	Moderate/Low	Class C
3.3	Rule 3	Yes	High	Moderate/Low	Class C
3.4	Rule 3	Yes	High	Moderate/Low	Class C
3.5	Rule 3	Yes	High	Moderate/Low	Class C
4.1	Rule 4	Yes	High	Moderate/Low	Class C
4.2	Rule 4	Yes	Moderate/Low	Low	Class B
5.1	Rule 5	Yes	Low	Low	Class A

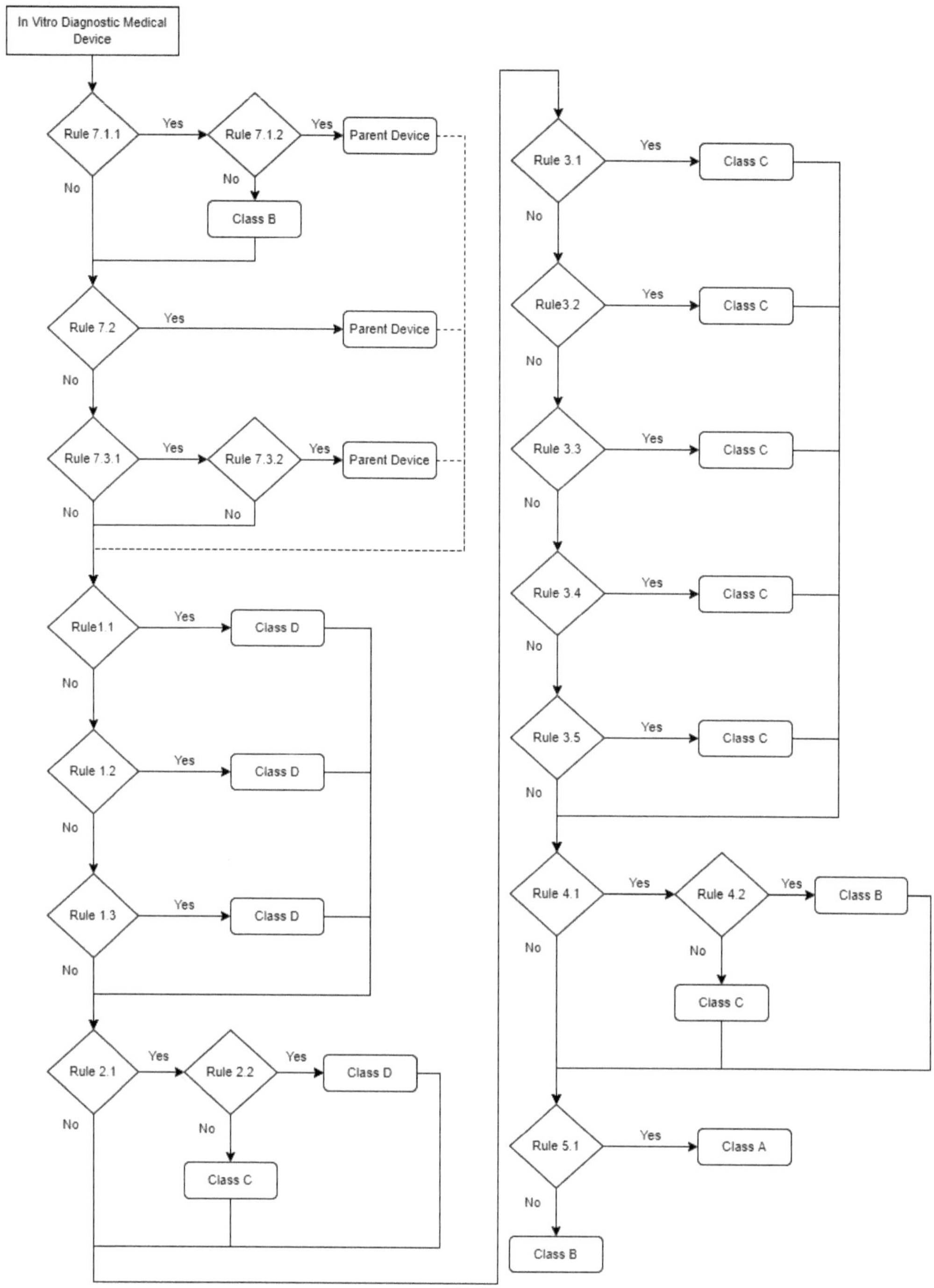

Figure 3. IVDR Decision Tree with Rules, classification rules and device risks corresponding to the IVDR Classification Rules and criteria in Table 2. The final classification given to a device consisted of Class A, B, C or D.

4.2. Classification Results from the IVDR Decision Tree, Build Using the RBDT-1C Algorithm

The IVDR decision tree was tested using 55 example IVD devices with known classifications according to the IVDR. This list of devices, which included devices from all of the classification and implementing rules, was compiled from multiple sources including: Medicines and Healthcare products Regulatory Agency (MHRA) In Vitro Diagnostic Directive (IVDD) guidance document [32]; British Standards Institute (BSI) classification of IVD devices whitepaper and infographic [33,34]; Australian Therapeutic Goods Administration (TGA) classification of IVD devices [35]; Food and Drug Administration (FDA) IVD device examples for human genetic tests and companion diagnostics [36,37]; and an International Society of Blood Transfusion (ISBT) report on human blood grouping systems [38]. Each of the devices was classified using the IVDR decision tree, independently by three experts, which consisted of medical and regulatory specialists. The independent classifications were assigned using the IVDR decision tree without prior knowledge of the known classifications. The results were discussed collaboratively to highlight different interpretations of the rules within Annex VIII of the IVDR and human error within the results. Ethical approval was in place (R63968/RE001) when collecting this data. Classification of 52 of the 55 device examples were agreed by all members of the research team, and these results were compared to the ground truth classifications. The decision tree was shown to accurately classify all IVD devices where there was no ambiguity in interpretation of the rules in the IVDR. Classification was not agreed for three devices, which identified ambiguity and interpretation issues within the classification rules of the IVDR.

The process of design and testing of the RBDT-1C identified difficulties in interpretation of the legislation. These issues resulted from differences between legal and technical semantics, and consisted of: ambiguity in terms used to distinguish between different device classifications; interpretation issues within classification rules; vagueness in healthcare domain specific vocabulary; and a lack of coherence between cross-referenced regulatory texts.

5. Discussion

The approach of converting rules within legislation into a rule-based decision tree identified specific interpretation issues within the IVDR. These interpretation issues were identified when implemented in January 2020, many of which were independently clarified in guidance documents that were published in November 2020 [39] by the designated European Medical Device Coordination Group (MDCG). This demonstrates that the introduced RBDT-1C algorithm can help legislators identify areas where guidance and further clarification is required. As well as this, digitizing the IVDR legislation identified cross referencing errors in the IVDR (and Medical Device Regulation), which have already been recognized and amended by the European Commission.

The medical device regulation field is highly leveraged on experience, and this process gives knowledge and insight into the regulation that usually takes years to capture, which is especially important for new and recent legislations. It was shown that this kind of algorithm could help IVD device manufacturers conform to, and better understand the IVDR legislation as the logic-based evaluation allows understanding of the interaction between the classification rules within the legislation, and provides insight into how users navigate the complex regulation. This approach can be used for other legislation and is not limited to the medical device industry. The conversion of legislation into a digital decision-making tool can also aid engineering management decisions when considering regulatory strategy within organizations.

It should be noted that these algorithms should keep in line with any rule changes. Outdated representation of the rules within an algorithm can lead to incorrect outcomes. Applying these kinds of algorithms should therefore also include a clear process on how to keep them up-to-date. This is particularly important when these algorithms are integrated within larger digital platforms. The digitization of legal frameworks can also take into account these considerations in order to allow software engineers to build more robust systems.

In general, resources that aid regulatory strategy can increase the understanding of legislation for all stakeholders within an industry. This can help overcome the barrier to innovation, which is often presented by regulation. While this is not limited to the medical device industry, this research example showed that digital solutions can be built to help users navigate the regulations. Nonetheless, this rule-based methodology is limited to legislation with structured and well defined rules, as the attributes need to be clearly defined according to the legislation in order to allow for accurate classification based on a rule-based decision tree. An additional, limitation of this method is the interpretation issues within the legislation, as ambiguity within the legislation resulted in divergent classification when using the IVDR decision tree. However, we do believe that highlighting and identifying these challenges can also lead to further additional clarification and guidance from the legislators. This could in turn result in a better understanding of regulations themselves. The introduced RBDT-1C algorithm provides a valuable tool within the field of regulatory science and engineering management.

Supplementary Materials: The following supporting information can be downloaded at: https://www.mdpi.com/article/10.3390/prosthesis4010012/s1, RBDT-1C mathematical supporting documentation.

Author Contributions: Conceptualization, J.H.M.B.; methodology, H.R.F.M. and J.H.M.B.; data collection, H.R.F.M. and R.H.; formal analysis, H.R.F.M.; data curation, H.R.F.M. and R.H.; writing—original draft preparation, H.R.F.M. and J.H.M.B.; writing—review and editing, H.R.F.M., A.C., R.H. and J.H.M.B.; visualization, H.R.F.M.; supervision, J.H.M.B.; project administration, H.R.F.M., R.H. and J.H.M.B.; funding acquisition, J.H.M.B. and R.H. All authors have read and agreed to the published version of the manuscript.

Funding: This research was funded by the European Institute of Innovation and Technology (EIT) Health [210692, EIT Health Education Portfolio, 2021] and supported by the National Institute for Health Research (NIHR) Oxford Biomedical Research Centre (BRC) [OxBRC4] and The Engineering and Physical Sciences Research Council (EPSRC) [EP/R511742/1].

Institutional Review Board Statement: The study was conducted in accordance with the Declaration of Helsinki, and approved by the Medical Sciences Interdivisional Research Ethics Committee (IDREC) of Oxford University (R63968/RE001), 27 June 2019.

Data Availability Statement: Not applicable.

Acknowledgments: The authors would like to acknowledge Azad Hussain and Eleanor Hughes for classifying example in vitro diagnostic (IVD) devices using the IVDR decision tree, and Oliver Lorton for implementing the designed tree for online testing.

Conflicts of Interest: The authors declare no conflict of interest. The funders had no role in the design of the study; in the collection, analyses, or interpretation of data; in the writing of the manuscript, or in the decision to publish the results.

References

1. King, A. Innovation from differentiation: Pollution control departments and innovation in the printed circuit industry. *IEEE Trans. Eng. Manag.* **1995**, *42*, 270–277. [CrossRef]
2. Wagner, M.; Bachor, V.; Ngai, E.W.T. Engineering and technology management for sustainable business development: Introductory remarks on the role of technology and regulation. *J. Eng. Technol. Manag.* **2014**, *34*, 1–8. [CrossRef]
3. Gerstenfeld, A. Government regulation effects on the direction of innovation: A focus on performance standards. *IEEE Trans. Eng. Manag.* **1977**, *EM-24*, 82–86. [CrossRef]
4. De Jesus Pacheco, D.A.; Caten, C.S.T.; Jung, C.F.; Navas, H.V.G.; Cruz-Machado, V.A. Eco-innovation determinants in manufacturing SMEs from emerging markets: Systematic literature review and challenges. *J. Eng. Technol. Manag.* **2018**, *48*, 44–63. [CrossRef]
5. Arnould, A.; Hendricusdottir, R.; Bergmann, J. The complexity of medical device regulations has increased, as assessed through data-driven techniques. *Prosthesis* **2021**, *3*, 314–330. [CrossRef]
6. Ettlie, J.E.; Bridges, W.P. Environmental uncertainty and organizational technology policy. *IEEE Trans. Eng. Manag.* **1982**, *EM-29*, 2–10. [CrossRef]
7. Sanchez, C.M.; McKinley, W. Environmental regulatory influence and product innovation: The contingency effects of organizational characteristics. *J. Eng. Technol. Manag.* **1998**, *15*, 257–278. [CrossRef]

8. Çanakoğlu, E.; Erzurumlu, S.S.; Erzurumlu, Y.O. How data-driven entrepreneur analyzes imperfect information for business opportunity evaluation. *IEEE Trans. Eng. Manag.* **2018**, *65*, 604–617. [CrossRef]
9. Vries, M.J.D. Translating customer requirements into technical specifications. In *Handbook of the Philosophy of Science, Philosophy of Technology and Engineering Sciences*; North-Holland: Amsterdam, The Netherlands, 2009; pp. 489–512.
10. Tiersma, P.M. The legal lexicon. In *Legal Language*; University of Chicago Press: Chicago, IL, USA, 1999.
11. Otto, P.N.; Anton, A.I. Addressing legal requirements in requirements engineering. In Proceedings of the 15th IEEE International Requirements Engineering Conference (RE 2007), Delhi, India, 15–19 October 2007; pp. 5–14. [CrossRef]
12. Sergot, M.J.; Sadri, F.; Kowalski, R.A.; Kriwaczek, F.; Hammond, P.; Cory, H.T. The British nationality act as a logic program. *Commun. ACM* **1986**, *29*, 370–386. [CrossRef]
13. Kowalski, R.A. Legislation as logic programs. In *Informatics and the Foundations of Legal Reasoning, Logic Programming in Action Lecture Notes in Computer Science*; Springer: Dordrecht, The Netherlands, 1995; pp. 235–256.
14. Leith, P. Fundamental errors in legal logic programming. *Comput. J.* **1986**, *29*, 545–552. [CrossRef]
15. Massacci, F.; Prest, M.; Zannone, N. Using a security requirements engineering methodology in practice: The compliance with the Italian data protection legislation. *Comput. Stand. Interfaces* **2005**, *27*, 445–455. [CrossRef]
16. Zhang, N.; Bodorik, P.; Jutla, D.N. Compliance of privacy policies with legal regulations—Compliance of privacy policies with Canadian PIPEDA. In Proceedings of the Second International Conference on e-Business, Barcelona, Spain, 28–31 July 2007.
17. Bergmann, J.; Hendricusdottir, R.; Lee, R. Regulatory navigation: A digital tool to understand medical device classification pathways. *Compr. Biotechnol.* **2019**, *5*, 167–172.
18. Quinlan, J.R. Induction of decision trees. *Mach. Learn.* **1986**, *1*, 81–106. [CrossRef]
19. Imam, I.F.; Michalski, R.S. Learning decision trees from decision rules: A method and initial results from a comparative study. *J. Intell. Inf. Syst.* **1993**, *2*, 279–304. [CrossRef]
20. Abdelhalim, A.; Traore, I.; Nakkabi, Y. Creating decision trees from rules using RBDT-1. *Comput. Intell.* **2014**, *32*, 216–239. [CrossRef]
21. Kotsiantis, S.B. Decision trees: A recent overview. *Artif. Intell. Rev.* **2011**, *39*, 261–283. [CrossRef]
22. Michalski, R.S.; Imam, I.F. Learning problem-oriented decision structures from decision rules: The AQDT-2 system. In *Lecture Notes in Computer Science Methodologies for Intelligent Systems*; Springer: Berlin/Heidelberg, Germany, 1994; pp. 416–426.
23. Trevino, M. 8 Key Changes to Understand in the New European MDR and IVDR. Med. Device Online, September 2018. Available online: https://www.meddeviceonline.com/doc/key-changes-to-understand-in-the-new-european-mdr-and-ivdr-0001 (accessed on 13 January 2020).
24. Ceross, A.; Bergmann, J. Evaluating the presence of software-as-a-medical-device in the australian therapeutic goods register. *Prosthesis* **2021**, *3*, 221–228. [CrossRef]
25. Obelis, Notified Bodies: IVDR. Obelis, 05 April 2019. Available online: https://www.obelis.net/notified-bodies-in-vitro-diagnostic-medical-devices/ (accessed on 13 January 2020).
26. Notified Bodies—Post Brexit. BSIF, 01 January 1967. Available online: https://www.bsif.co.uk/notified-bodies-post-brexit/ (accessed on 27 April 2020).
27. European Commission. Regulation (EU) 2017/746 of the European Parliament and of the Council of 5 April 2017 on In Vitro Diagnostic Medical Devices and Repealing Directive 98/79/EC and Commission Decision 2010/227/E. 2017.
28. Lietz, P. Research into questionnaire design: A summary of the literature. *Int. J. Mark. Res.* **2010**, *52*, 249–272. [CrossRef]
29. Holbrook, A.; Cho, Y.I.; Johnson, T. The impact of question and respondent characteristics on comprehension and mapping difficulties. *Public Opin. Q.* **2006**, *70*, 565–595. [CrossRef]
30. Foddy, W. The limitations of human memory. In *Constructing Questions for Interviews and Questionnaires: Theory and Practice in Social Research, Cambridge*; Cambridge University Press: Cambridge, UK, 1993; pp. 90–100. [CrossRef]
31. Lee, J.; Yoon, W.; Kim, S.; Kim, D.; Kim, S.; So, C.H.; Kang, J. BioBERT: A pre-trained biomedical language representation model for biomedical text mining. *Bioinformatics* **2019**, *36*, 1234–1240. [CrossRef]
32. MDCG 2019-11 Guidance on Qualification and Classification of Software in Regulation (EU) 2017/745—MDR and Regulation (EU) 2017/746—IVDR. 2019.
33. British Standards Institution. IVDR Classification and Conformity Assessment. 2017.
34. British Standards Institute. Classification of IVDs Infographic. 2020.
35. Australian Government Department of Health. IVD Classification Examples. Therapeutic Goods Administration (TGA), 06-December 2015. Available online: https://www.tga.gov.au/book-page/ivd-classification-examples#r14eg (accessed on 16 February 2020).
36. Center for Devices; Radiological Health. Nucleic Acid Based Tests. U.S. Food and Drug Administration. Available online: https://www.fda.gov/medical-devices/vitro-diagnostics/nucleic-acid-based-tests (accessed on 19 February 2020).
37. Center for Devices; Radiological Health. List of Cleared or Approved Companion Diagnostic Devices. U.S. Food and Drug Administration. Available online: https://www.fda.gov/medical-devices/vitro-diagnostics/list-cleared-or-approved-companion-diagnostic-devices-vitro-and-imaging-tools (accessed on 1 March 2020).

38. Daniels, G.L.; Fletcher, A.; Garratty, G.; Henry, S.; Jorgensen, J.; Judd, W.J.; Levene, C.; Lomas-Francis, C.; Moulds, J.J.; Moulds, J.M.; et al. Blood group terminology 2004: From the International Society of Blood Transfusion committee on terminology for red cell surface antigens. *Vox Sang.* **2004**, *87*, 304–316. [CrossRef]
39. Medical Device Coordination Group. MDCG 2020-16: Guidance on Classification Rules for In Vitro Diagnostic Medical Devices under Regulation (EU) 2017/746. 2020.

prosthesis

MDPI

Communication

Medical Device Regulation from a Health Service Provider's Perspective

Man Ting Kwong [1], David Stell [1,*] and Emmanuel Akinluyi [1,2]

[1] Clinical Engineering, Department of Medical Physics, Guy's and St Thomas' NHS Foundation Trust, London SE1 9RT, UK; emily.kwong@gstt.nhs.uk (M.T.K.); Emmanuel.Akinluyi@gstt.nhs.uk (E.A.)

[2] School of Biomedical Engineering & Imaging Sciences, King's College London, London SE1 7EH, UK

* Correspondence: david.stell@gstt.nhs.uk

Abstract: Unfamiliarity with medical device regulations can sometimes be a barrier to deploying technology in a clinical setting for researchers and innovators. Health service providers recognise that innovation can happen within smaller organisations, where regulatory support may be limited. This article sets out to increase transparency and outline key considerations on medical device regulations from a UK healthcare provider's perspective. The framework used by Guy's and St Thomas' NHS Foundation Trust (GSTFT) for assessing research devices is presented to give an overview of the routes that R&D medical devices take to enter a clinical setting. Furthermore, current trends on research studies involving medical devices were extracted from the GSTFT internal R&D database and presented as the following categories (i) commercial vs. non-commercial, (ii) assessment type and (iii) software vs. non-software. New medical devices legislation will be introduced within the UK in July 2023. It is anticipated regulating software as a medical device may become more challenging for healthcare providers and device manufacturers alike. It is therefore important for different stakeholders involved to work together to ensure this does not become a barrier to innovation.

Keywords: medical device regulations; healthcare provider; software as medical device; technology translation; quality management system

Citation: Kwong, M.T.; Stell, D.; Akinluyi, E. Medical Device Regulation from a Health Service Provider's Perspective. *Prosthesis* **2021**, *3*, 261–266. https://doi.org/10.3390/prosthesis3030025

Academic Editor: Marco Cicciu

Received: 3 August 2021
Accepted: 8 September 2021
Published: 14 September 2021

1. Introduction

Innovation can often happen at the level of a research organization or by an early-stage innovator. Development of medical devices may require extensive clinical investigation and evaluation to ensure patient safety. Navigating the path towards clinical investigation and evaluation could pose a barrier to early-stage innovators and hence may impede innovation. The aim of this article is to give early-stage researchers and innovators an overview of the regulatory environment in the UK from a healthcare provider's perspective, and possible routes for R&D medical devices to enter testing in a clinical setting.

In the UK, medical devices are regulated under the Medical Devices Regulations 2002 (commonly referred to as the UK MDR 2002). This legislation implemented existing EU medical devices directives into UK law. The legislation required CE marking for all medical devices sold within the UK, to indicate compliance with EU legislation. The most widely applicable of the EU directives is the Medical Devices Directive (Directive 93/42/EEC, commonly referred to as the MDD) [1]. The legal requirements for medical device manufacturers are largely contained within the MDD, with some UK specific requirements contained within the UK Medical Device Regulations 2002 [2].

The MDD is no longer in force within the EU, having been replaced by the EU Medical Devices Regulation (2017/745, commonly referred to as the MDR) [3], on the 26 May 2021. New medical devices legislation will be introduced within the UK in July 2023, the contents of the new legislation are not known but may have similarities to the EU MDR.

Alongside the MDD, the UK Medical Devices Regulations 2002 implemented two other EU directives into UK law; the Active Implantable Medical Devices Directive

(AIMDD, 90/385/EEC) [4] and the In Vitro Diagnostic Medical Devices Directive (IVDD, 98/79/EEC) [5]. Within the UK, the legal requirements for these devices remain largely contained within the EU directives. Within the EU the IVDD remains in force. The AIMDD has now been withdrawn, with the requirements for active implantable medical devices now contained within the MDR.

2. Role as a Gatekeeper for Medical Devices

Healthcare providers in the UK have a statutory responsibility to ensure patient safety. Most health providers have a process for approving and accepting medical devices for local deployment. Clinical engineers will often act as gatekeepers, assessing and approving new medical devices/devices being used for research. As an example to help improve transparency, the framework used by the Guy's and St Thomas' NHS Foundation Trust (GSTFT) for assessing research devices is presented below.

There are several possible routes for R&D devices to enter a health service, these are broadly categorized into: (i) Commercial R&D, (ii) Non-commercial R&D and (iii) Non-R&D. Each of these categories will be described in more detail in the rest of this section. It is important to note that ethics approval is required for medical device research involving patients.

2.1. Commercial R&D

The sponsor of commercial R&D studies is typically a commercial body, usually either a medical devices manufacturer or a pharmaceutical company. Clinical engineers are involved in the assessment of any study which involves medical devices, irrespective of who the sponsor is.

Studies run by pharmaceutical companies may involve medical devices. These studies do not typically assess the medical devices themselves; these devices are ancillary, for instance diagnostic devices may be used to assess treatment outcomes. In these cases, the devices used are post-market, with pre-existing regulatory approval. Clinical engineers' involvement in the assessment of these studies is typically limited to checking that devices do carry the necessary regulatory approval (i.e., UKCA or CE marking) and are used within their intended purpose.

Studies run by medical device manufacturers are more likely to assess the performance of medical devices themselves. These studies may involve pre-market or post-market devices. Even studies of pre-market devices typically involve devices whose designs are fairly mature.

Trusts typically charge more for commercial studies than non-commercial ones and retain less ownership of the data produced by the study.

2.2. Non-Commercial R&D

The sponsor of non-commercial studies is typically a non-commercial body, often the NHS Trust itself, or a university. These studies are very diverse. Non-commercial R&D studies which have a medical device as the subject of the study can involve early–stage technologies whose designs are less mature than those typical of commercial pre-market evaluations.

Trusts typically charge less for non-commercial studies than commercial ones but will have stricter contractual limitations governing what data can be shared with the sponsor.

2.3. Non R&D

In some cases, new medical devices, medical device modifications, or systems of medical devices are not for research purposes, they are to meet an identified clinical need within the Trust. Devices we have seen within GSTFT include fixation devices to minimise patient movement during imaging, and incubator trolleys mounting multiple devices which require mitigations to assure electrical safety.

3. Internal Assessment Considerations

Clinical engineers involved in the assessment of medical devices for deployment within their local institution must consider several factors. Their ultimate responsibilities are to ensure that any deployment is safe and does not adversely affect the patient care delivered by their organisation. However, the assessment questions are also driven by the regulatory route used for the deployment, and local policies and norms.

Regulatory approval provides a high degree of assurance of device safety and performance. A correspondingly light assessment is therefore required for these devices.

Where a study has been notified to the Medicines and Healthcare products Regulatory Agency (MHRA), and they have indicated that they have no concerns, this too provides a high degree of assurance and reduces the scope of the assessment required from local clinical engineers.

No such assurance exists for medical devices which do not have regulatory approval, and where studies have not been notified to the MHRA. Deployment of these devices is permitted in a limited range of specific circumstances. Assessments of these deployments include detailed technical assessments, together with a consideration of the legal deployment route.

4. Approval and Assessment Type

Based on the above factors mentioned, the decision pathway to ensure patient safety could be categorised into the following assessment types: (i) Trust owned medical devices, (ii) UKCA/CE marked devices, (ii) registered clinical investigations, (iii) basic safety assessments and (iv) Health Institute Exemption.

4.1. Trust Owned Medical Devices

Some researchers wish to use medical devices which are already in clinical use at a Trust for their research. Provided that the devices are not being modified or used for a novel purpose, there are no regulatory barriers to this use.

4.2. UKCA/CE Marked Medical Devices

It is very common for research studies to involve use of medical devices which are UKCA/CE marked and available for purchase. There are no further regulatory requirements to this type of use. However, individual Trusts may have local governance arrangements which determine which medical devices may be introduced to the Trust. New devices may have to meet local governance requirements before they may be introduced to clinical areas.

4.3. Registered Clinical Investigations

To demonstrate that their device conforms to the regulatory requirements, manufacturers must sometimes run clinical studies known as *clinical investigations*. In many respects these are the medical devices equivalent of a clinical trial of an investigation medicinal product (CTIMP). Within the UK, the MHRA must be notified of any clinical investigation and have the opportunity to object before it may proceed.

Clinical investigations are typically run as commercial R&D projects within NHS Trusts. Once the MHRA have indicated that they do not object, Trusts are usually happy for the device to be used with their patients, subject to the usual R&D governance requirements.

This is a common but resource intensive route for devices to enter a NHS Trust, the manufacturer may act as the sponsor of the clinical investigation. This will commonly involve dedicated regulatory experts and clinical trials units. Securing funding for this step is a common barrier for devices undergoing the innovation translation process.

4.4. Basic Safety Assessments

In some cases it is possible for an investigational medical device to be used within a clinical study without being used for any medical purpose. For example, a diagnostic device may be used to make clinical measurements which are used for research purposes (e.g., for comparison with measurements made by a gold standard technology) but are not used for any clinical purpose (i.e., they are not used by clinicians to make decisions about patients' care).

Existing medical devices legislation applies only to devices intended by their manufacturer to be used for medical purposes. Investigational devices which are not intended to be used in this way do not need to comply with these legislative requirements. Healthcare providers are nonetheless required to ensure that such devices are safe for use with patients and will often wish to perform a "basic safety assessment" for these devices.

Clinical medical devices research conducted under a basic safety assessment can be a good option for studies involving medical devices whose design is not yet mature, and which can be meaningfully assessed without impacting patients' clinical care. However, data from these studies cannot typically be used to demonstrate devices' safety or performance as part of a formal conformity assessment process. A clinical investigation, notified to the MHRA, is required where research outputs are intended to be used in this way.

Basic safety assessments are also sometimes used to ensure the safety of non-medical devices which are introduced to clinical areas as part of a clinical study.

4.5. *Health Institute Exemption*

Within the EU, the MDR has introduced new requirements for healthcare institutions who wish to manufacture and use medical devices internally. These requirements are more demanding for healthcare providers than those under the previous legislation. There are requirements to be more structured and transparent, such as for health institution to draw up publicly available declaration that the devices meet the general safety and performance requirements. Most notably in the MDR there is now a specific requirement for an appropriate quality management systems. This will require that health institutions adapt their existing quality management system as they adopt new technology that is developed and manufactured in-house.

It is important to note that at the time of writing, the MDR does not apply in the UK. However, it is prudent for researchers developing medical devices to consider the quality management system requirements early. There is a need to make such systems more widely available to research institutions and harmonised with partnering healthcare providers.

5. Current Trends of R&D Devices

GSTFT is the largest NHS Trust in the UK. It has a well-established research partnership with King's College London, as part of the King's Health Partnership. It is therefore a very research active centre. The following summarises some key statistics extracted from our internal database of R&D device research activity between 2019 and 2021.

Table 1 presents the proportion of commercial and non-commercial clinical research and Table 2 presents the proportions of the different types of assessments conducted in 2019. Figures from January–December 2019 were chosen, as they represent the 12 months leading up to the start of the COVID-19 pandemic and may be informative of the amount of research activity that would occur when there were no disruption of the service.

Table 1. Percentage and number of studies GSTFT received in 2019 via commercial and non-commercial routes.

Route to Healthcare Providers	% (Number of Studies)
Commercial	33% (27)
Non-commercial	60% (49)
Non-R & D	7% (6)

Table 2. Percentage and number of studies GSTFT received in 2019 under different assessment types.

Assessment Type	% (Number of Studies)
Trust owned medical devices	2% (2)
CE marked medical devices	45% (37)
Registered clinical investigations	4% (3)
Basic safety assessments	31% (21)
Health Institute Exemption	7% (5)

The MDD included software under the "active device" category after an amendment made in 2007, software as a medical device is covered and can be classified. New legislation outlined in the MDR, however, prescribed a more robust and specific approach to regulating software as medical devices.

This change could present a particular challenge for healthcare providers because there are likely to be many non-R&D software algorithms (e.g., spreadsheet algorithms) which require more rigorous governance under the new legislation. This type of non-R&D software algorithm will likely fall in the HIE category. The MDR has many more requirements for healthcare providers who wish to use the HIE than does the MDD.

Table 3 presents the number of research projects that involved software vs. non-software. The proportion of studies classified as software was low in 2019. If new medical devices legislation in the UK mirrors the EU MDR then the number of medical device software assessments are anticipated to rise as non-R&D algorithms are brought into compliance with the new legislation.

Table 3. Percentage and number of studies GSTFT received in 2019 grouped under software vs. non-software.

Software vs. Non-Software	% (Number of Studies)
Software	4% (4)
Non-software	96% (104)

Figure 1 shows the number of studies received via the commercial and non-commercial routes since Jan-19. There is generally less commercial research activity than non-commercial, but non-commercial research activity is the more variable. All research activity stopped during the first peak of COVID-19 pandemic and levels have been very erratic since then, particularly for non-commercial research.

6. Discussion

Healthcare providers are aware that innovation is key to improving healthcare. Medical device innovation may occur at an innovator or research organization level, where routes to clinical deployment for research may pose a hurdle. This article has discussed the various pathways which exist within an NHS Trust and has discussed the range of research types supported by healthcare providers.

With the introduction of new medical device legislation in Europe, and new legislation also on the horizon for the UK, this article aims to increase transparency over healthcare providers' considerations when approving a piece of medical device for clinical studies. The largest change in terms of the device approval from a healthcare provider's perspective will likely be the increased requirements for deployments under the HIE. This imposes on the healthcare provider a similar responsibility to a commercial manufacturer, such as a more specific process governance with a quality management system. This may require adaptation of processes when introducing a newly developed in-house technology. From manufacturers' points of view, it is likely that the tighter classification rules and conformity assessment routes will pose a challenge. Transparency and cross-learning here is important as Trusts and healthcare providers will each have new responsibilities to ensure all software complies with the new legislation requirements.

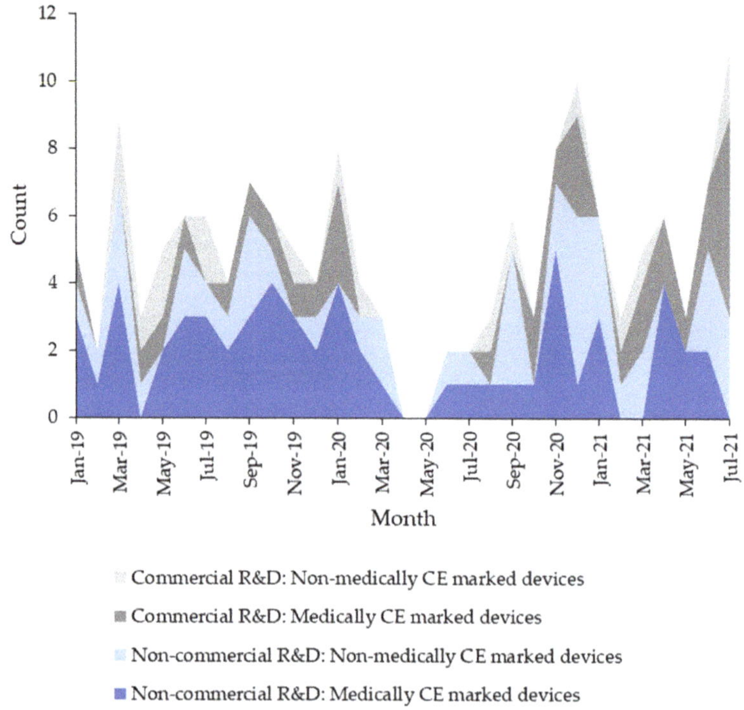

Figure 1. Volume of research studies involving medical devices assessed by GSTFT received in 2019–2021.

The volume of clinical studies is likely to rise in the future as the healthcare demand created by the COVID-19 pandemic falls. Research activity is returning to pre-pandemic levels and may soon exceed historic levels as Trusts begin clearing the backlog of research studies delayed by the pandemic, and with the additional research opportunities created by the pandemic.

Author Contributions: Conceptualization, M.T.K., D.S. and E.A.; methodology, M.T.K. and D.S.; formal analysis, M.T.K. and D.S.; data curation, D.S.; writing—original draft preparation, M.T.K.; writing—review and editing, D.S. and E.A.; All authors have read and agreed to the published version of the manuscript.

Funding: This research received no external funding.

Institutional Review Board Statement: Not applicable.

Informed Consent Statement: Not applicable.

Conflicts of Interest: The authors declare no conflict of interest.

References

1. European Council Directive 93/42/EEC of 14 June 1993 Concerning Medical Devices. *Off. J. Eur. Union L* **1993**, *169*, 1–43.
2. Secretary of State. The Medical Devices Regulations 2002, No. 618, UK Statutory Instruments.
3. European Union. Regulations (EU) 2017/745 of the European parliament and of the council of 5 April 2017 on medical devices. *Off. J. Eur. Union L* **2017**, *117*, 1–175.
4. Council Directive 90/385/EEC of 20 June 1990 on the Approximation of the Laws of the Member States Relating to Active Implantable Medical Devices. *Off. J. Eur. Union L* **1990**, *189*, 17–36.
5. Directive 98/79/EC of the European Parliament and of the Council of 27 October 1998 on In Vitro Diagnostic Medical Devices. *Off. J. Eur. Union L* **1998**, *331*, 1–37.

prosthesis

Review

Evaluating the Presence of Software-as-a-Medical-Device in the Australian Therapeutic Goods Register

Aaron Ceross * and Jeroen Bergmann

Department of Engineering Science, University of Oxford, Oxford OX1 3PJ, UK; jeroen.bergmann@eng.ox.ac.uk
* Correspondence: aaron.ceross@eng.ox.ac.uk; Tel.: +44-1865-273000

Abstract: In recent years, medical device regulatory bodies have recognised software-as-a-medical-device (SaMD) as a distinct subgroup of devices. The field of SaMD has been rapidly evolving and encompasses a range of different digital solutions. Many organisations have now started to look into digital healthcare, as a way to solve key global challenges. However, there remains uncertainty regarding how many of these SaMD products are entering the market and to what extent these systems achieve a desired level of general safety once they are in the market. In this study, we utilise data collected from publicly available databases. The data are evaluated for trends and a descriptive analysis is performed of the recall and adverse events associated specifically with SaMD. We find that there is a significant positive trend ($p < 0.05$) of SaMD registrations, although the number of SaMD registrations remains relative low compared to non-SaMD. This rise in SaMD registrations coincides with increasing levels of recalls and adverse events. More importantly, it becomes apparent that adverse events notification is not yet fit for purpose with regards to SaMD.

Keywords: medical technology; digital healthcare; data science; regulations; machine learning; software

Citation: Ceross, A.; Bergmann, J. Evaluating the Presence of Software-as-a-Medical-Device in the Australian Therapeutic Goods Register. *Prosthesis* **2021**, 3, 221–228. https://doi.org/10.3390/prosthesis3030022

Academic Editor: Marco Cicciu

Received: 24 June 2021
Accepted: 20 August 2021
Published: 24 August 2021

Publisher's Note: MDPI stays neutral with regard to jurisdictional claims in published maps and institutional affiliations.

1. Introduction

Although the inclusion of software into medical devices is a long established practice, stand alone software offerings that provide diagnostic and therapeutic functions is relative new. Nonetheless, software applications are an increasingly fundamental feature of clinical practice and regulators around the world have responded to recognise this rise in interest within this area. Certain regulators, such as the one in Australia, have even made data available that can be used for further analysis. The use of medical devices in Australia is overseen by the Therapeutic Goods Authority (TGA). Consideration of software in medical devices has long been a concern for the TGA, which had evaluated means by which to address software through its essential principles as early as 2001 [1]. In 2020, the TGA released a study on the harms caused by software [2], outlining recalls and adverse events related to software through a literature review. The TGA report did not distinguish in its analysis between medical devices that have software and software as a stand alone medical device.

The distinction between devices that incorporate software and "software-as-a-medical-device" (SaMD) is important, as these now represent two distinct device types within regulation. SaMD are those entirely composed of software without any additional hardware. In other words, the software program itself is the medical device. Since 2013, there has been international consensus through the International Medical Device Regulators Forum (of which Australia is a member) on the definition of SaMD [3]. Although regulatory legislation now incorporates the definition, there remain unanswered questions about how the safety of such devices can be ensured. There has been work conducted related to recall and adverse events for medical devices that involve software using data from the FDA in the United States (e.g., [4,5]). However, we have been unable to find any quantitative analysis specific to SaMD in any jurisdiction after an extensive review of the available literature. Further to this, there has been no empirical treatment of reported risks associated with SaMD in Australia, or any other jurisdiction. To the best of our knowledge this work represents the first quantitative analysis of SaMD within any regulatory framework. In this work, we focus on the following three research questions; (1) Has SaMD registrations been growing in Australia and what is the origin (domestic or foreign)? (2) What types of recall events are associated with SaMD? (3) What are the types of adverse events associated with SaMD? We conclude with a discussion on results and the nature of regulatory actions for SaMD.

2. Data and Methods

2.1. Data Collection

For this study, we relied on the TGA's publicly accessible databases relating to (a) the registration of medical devices, (b) the recall of the devices, and (c) adverse events associated with the device. We utilised these databases in order to map SaMD development in Australia, with a particular focus on risk of harms that SaMD might pose. Below is a description of databases that were used in this study:

- The Australian Register of Therapeutic Goods (ARTG), which is a publicly accessible database of medical devices that are allowed on the market in Australia. An entry in the ARTG contains the device's description, global medical device number (GMDN), manufacturer, sponsor, classification, and date of registration [6]. A GMDN provides a broad description of a device type and consists of a unique number. The TGA's website states that, as of October 2019, there were over 90,000 registrations on the register, which includes medicines, as well as medical devices. We extract ARTG entries resulting in 60,806 device records, comprised of 7029 unique device types (based on their GMDN), spanning a period of 18 years, from 26 November 2002 to 24 June 2020;
- The System for Australian Recall Action (SARA) is a searchable, public database containing product recall action notifications. These notifications are divided into (i) recalls; (ii) recalls for product correction; and (iii) hazard alerts. We extracted 4746 notifications, over the period between 2 July 2012 to 24 June 2020;
- The Database of Adverse Event Notifications (DAEN) records the types of harms caused by medical devices. Within this database the string 'software' was searched, which returned 68 events, of which there are 43 events that could be cross-referenced to the ARTG.

All the above databases are publicly available on the internet. In order to extract the data, a webscraper was developed for the ARTG and SARA. The webscraper interfaced with the search box provided by each of these databases. We searched only for medical devices, excluding medicines and other therapies. The webscraper downloaded the returned documents in the PDF format. The data were then extracted by transforming the PDF into plain text and extracting the information from the records..

2.2. Analysis Methodology

For the first research question, we undertook a trend analysis, evaluating whether SaMD has a trend and, if so, what the nature of that trend was. In order to make this determination, we utilise the Mann–Kendall test, which determines whether a time series exhibits a monotonic upward or downward trend. In relation to second question, a visualisation of recalls in SARA was created and the nature of those recalls related to SaMD was explored. Finally, the last question was investigated by obtaining the description of the types of adverse events associated with software generally, and the presence of SaMD in the dataset was determined. We developed the webscraper and conducted our analysis using version 4.1 of the R programming language.

3. Results

3.1. Registration of Software-as-a-Medical-Device

For all devices containing any kind of software, the ARTG dataset lists 971 devices, which represents 1.6% of the entire register of therapeutic goods. Of these, 736 were identified as SaMD with the remaining 235 being devices that incorporated software to some degree. We analysed the trend in SaMD registrations by finding the values related to the monthly registrations of SaMD. Figure 1 illustrates a pronounced increase in SaMD over time when compared to software that was incorporated into medical devices. The Mann–Kendall test showed that a significant positive trend for SaMD ($p < 0.05$) existed in the data. No significant trend was found for software that was incorporated into medical devices.

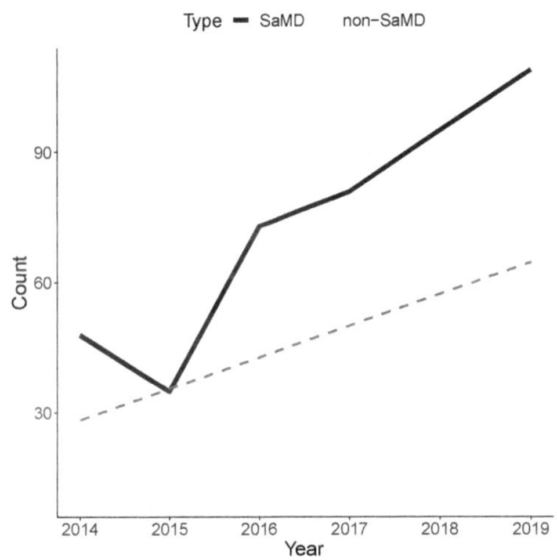

Figure 1. Times series comparison of software-as-a-medical-device (SaMD) and "software that incorporated into a medical device" (non-SaMD) based on the registrations in the ARTG over a period ranging from 2014–2019 (n = 558). The dotted red line represent the lowess regression smoothing of all the time series data.

Of the total amount of registered devices, foreign (non-Australian) products made up 92% with the remaining products being domestic (8%). Overall devices from either the USA and the EU constitute a total of 66.6% of the Australian market. However, there is a stronger domestic representation when SaMD is considered in its own right. Australian development of SaMD has a positive trend that is significant ($p < 0.05$) when tested with the Mann–Kendall test, which determines whether or not there is a linear monotonic trend, as illustrated in Figure 2.

The legislation provides a system of categorisation of medical devices based on the potential risk of harm to the user. Assessment for risk consideration include: (i) the intended use of the device, (ii) location on the body during use, (iii) invasiveness into the body, as well as (iv) duration of use. The Therapeutic Goods Act 1989, along with associated subsidiary legislation, regulates the use of medical devices in Australia. This legislation provides a classification system to assess the risk that devices can pose and provides the means to categorise these devices according to risk. There are three main categories for medical devices: (i) Class I for low risk, (ii) Class II for low to medium risk, (iii) Class IIB for medium to high risk, and (iv) Class III for the highest risk devices. The largest classification group for non-software devices and SaMDs is class I, while for other non-SaMD software, the largest group consist of those from class II. In Figure 3, we illustrate the division of risk classifications based on the type of device: (i) non-software; (ii) other software, meaning that the software is part of a physical device, and (iii) SaMD.

The proportions remain consistent across all device types, with Class I representing around half of the classifications (57%, 52%, and 51%, respectively). SaMD has a higher proportion of low-medium risk class devices Class IIA (32% compared to 21% for non-software devices and 28% for devices that include software) and lower proportion of high risk devices, Class III (2% compared to 9% for non-software, and 7% for software included in another device).

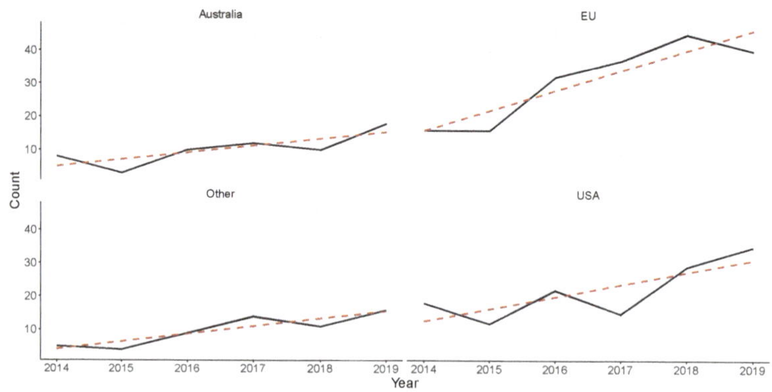

Figure 2. Time series data of software-as-a-medical-device (SaMD) for registrations, based on the region of origin as reported in the ARTG. The data covers a period between 2014 and 2019. Dotted red lines represent the lowess regression for each of the time series.

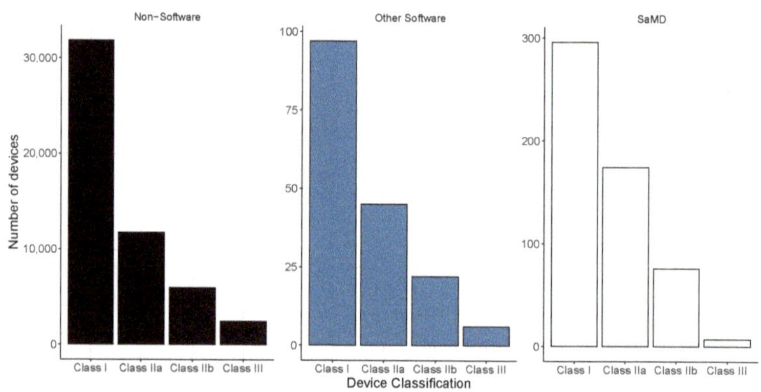

Figure 3. Comparison of device classification between software-as-a-medical-device (SaMD), other software and non-software. Please note the difference in scales for each of the figures.

3.2. Recall Actions

The System for Australian Recall Action (SARA) is a searchable, public database containing product recall action notifications. These notifications are divided into (i) recalls; (ii) recalls for product correction; and (iii) hazard alerts. It is important to note that entries into SARA are commenced by the product manufacturer or product sponsor with the TGA monitoring the recall action itself. A total of 4746 notifications were extracted, from between 2 July 2012 to 24 June 2020. When cross-referencing the ARTG data set with SARA, it resulted in 3597 recall events, obtained from 1900 unique devices, as shown in Table 1.

SARA provides the "level" at which the device is recalled, which relates to hospital, retail, wholesale, or consumer. The hospital setting accounted for over 90% of the recalls actions (Table 2), which might relate to the nature of the devices.

Table 1. Recall actions obtained from the System for Australian Recall Action (SARA) database for the period from July 2012 to June 2020.

Action	Count	Percentage
Recall for Product Correction	1396	38.81
Recall	1193	33.17
Product Defect Correction	891	24.77
Hazard Alert	104	2.89
Product Defect Alert	13	0.36

Table 2. The level at which recalls are made for all medical devices in System for Australian Recall Action (SARA).

Action Level	Count	Percentage
Hospital	3286	91.35
Retail	178	4.95
Consumer	118	3.28
Wholesale	15	0.42

There are 455 recall actions in SARA that mention software as the reason for the reported error, comprising 9.6% of all the reported recall events. A total of 72 software recalls were related to SaMD, constituting 16% of the software reported recalls and 2.3% of total recalls. All of these SaMD recalls come from 29 unique SaMD registrations, making up 3.9% of all registered SaMD that are in the ARTG. The connection between recall action, level and recall class is shown in Figure 4 using an alluvial plot to map them out for both SaMD and all others. The figure is set to a logarithmic scale in order to highlight the differences between SaMD and all other medical devices.

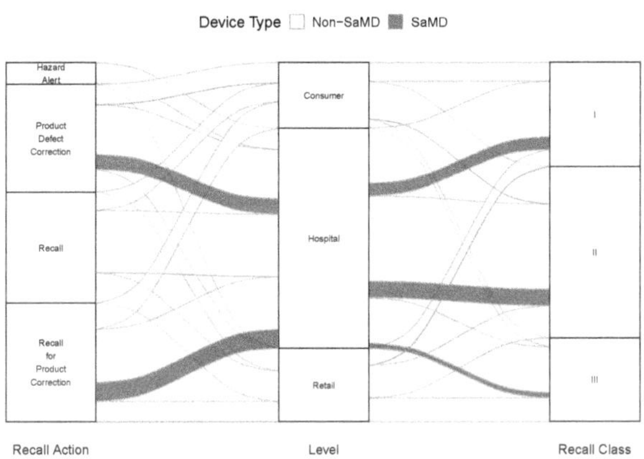

Figure 4. Alluvial diagram of recalls types, levels, and recall class in the SARA dataset. Data are plotted for SaMD and all other medical devices (Non-SaMD). The scale is logarithmic.

3.3. *Adverse Events*

The Database of Adverse Event Notifications (DAEN) records the types of harms caused by medical devices. We searched for the string 'software' in DAEN, which returned 68 events, of which there are 43 events that could be cross-referenced with data from the ARTG. This resulted in 17 devices, of which 7 were SaMD. Due to the nature of DAEN, we were unable to retrieve all reported events for all device types.

The event types in DAEN are categories according to the standard on medical device adverse event reporting [7]. The standard lists 20 different event types, 9 of which have been reported for software. According to the standard, all software errors fall under the broad event label of problems related to 'Computer Software', rather than providing specific software issue labels. As such, Figure 5 shows that the second-most frequent of the events in DAEN for software devices fall under this heading (n = 10, 26.3%) after usability issues (n = 11, 28.9%). Non-SaMD include a label describing software issues related to hardware control as well.

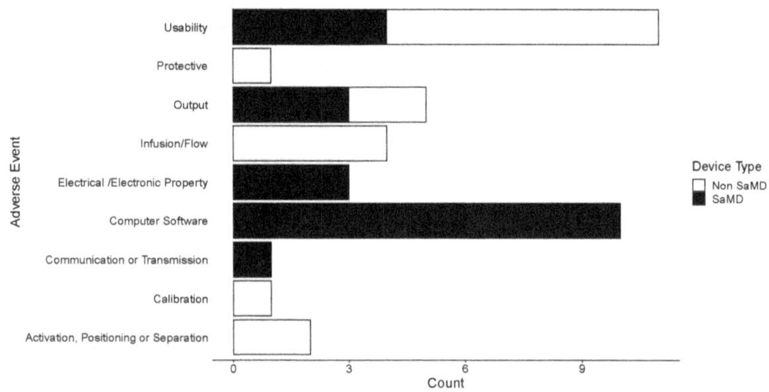

Figure 5. Number and type of adverse events for both SaMD and non-SaMD software.

The DAEN database provides injury outcomes of the reported events. Figure 6 shows that SaMD are associated with low levels of injury, but it should be noted that the "unknown" category is also high. This suggests that it is difficult to ascertain the exact level of injury from SaMD events.

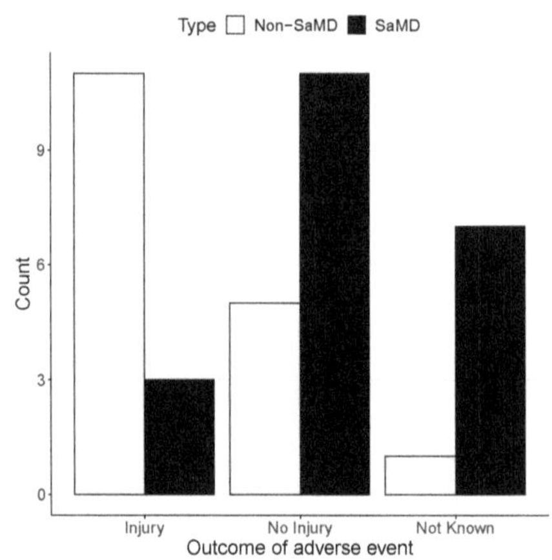

Figure 6. Outcomes of adverse event related to software in DAEN for both SaMD and non-SaMD software.

4. Discussion

The results demonstrate that SaMD (and software more generally) are growing facets of the medical device landscape, as shown based on the Australian data. Caution needs to be taken in case of generalising too much, as there might regional differences. Nonetheless, the suggestion seems to be in-line with the popularity of machine learning and artificial intelligence for healthcare within the research arena.

The outcomes also highlight that there are some challenges in evaluating the risks associated with SaMD, as the limited regulatory vocabulary for software defects prevents further granular analysis. The blunt categorisation of the ISO code "Computer Software Problem" in DAEN makes it hard to identify clear areas of improvement for the SaMD. It acts as an obstacle for the regulator, and those assessing regulatory data, as it prevents a better understanding of the types of events that occur for SaMD and reduces the potential for better risk management. Within the recall data, there are no fields specifying the type of problem that prompted the recall. This is not because a lack of vocabulary to describe software defects. This does exist, in, for example, the IEEE Standard for Classification of Software Anomalies [8]. The inclusion of more detailed software-specific recall reports and adverse events may facilitate the policy prioritisation for the regulator, which, in turn, will ensure that devices on the market reach an appropriate level of safety and performance.

The issue of applying appropriate standards for clinical safety are more acute for software than for non-software. McHugh et al. [9] already suggested that the rapid, iterative approach to software development may be in tension with the regulatory requirements for medical devices. Non-software device manufactures require specific fabrication infrastructure. This is in stark contrast to what happens in software testing. The costs associated with changing software are different compared to those associated with hardware manufacturing. For example, a software developer could quickly push an update to customers upon the discovery of a problem, whilst this is impossible for a (hardware) manufacture to do at scale. This rapid response option is not available to non-software devices, which must collect all defective items and fix the issue physically. Increasing the regulatory support offered to software developers during the early phases of research and development, as well as integrating this kind of support into, e.g., computer science courses, will potentially create more robust solutions for these regulatory issues in the future [10].

The difference in how risks need to be addressed by the manufacturers of products also relates to the origin of the product itself. The Australian market for medical devices relies heavily on foreign suppliers of medical devices. Herpin et al. [11] reports that the funding opportunities within the Australian biomedical industry may motivate innovators within the country to seek to establish manufacturing operations in larger markets for profit maximisation as the distance to these markets has an impact on costs. These economic dynamics seem to affect SaMD development less as there is no need for a factory to manufacture these devices, just access to skilled software developers. The lower initial cost for a more scalable product might account for the higher levels of Australian representation in SaMD.

In their work on safety and security of products in supply chain management, Marucheck et al. [12] found that recalls and adverse events of products, rather than regulation, may limit innovation in medical device development. Although Thirumalai and Sinha [13] showed that within the medical device industry, recalls costs are expected and they normally do not influence the market too much. As mentioned software might be better suited for ongoing improvements and this is also reflected in the fact that the for SaMD corrections are taking place if defects are detected.

We note that this study is limited by the amount of available data. The categorisation of 'SaMD' in this study was derived through the provided descriptions and the GMDN of the medical devices that were reported in the ARTG. There could be a potential for misclassification because of this. The results presented do not represent definitive evidence of trends, only an illustration of patterns available based on the data collected by the TGA. With the popularity of SaMD continuing to grow, along with the concerns that the TGA has with the regulation of these devices, it may be useful for records to indicate which devices are solely composed of software.

5. Conclusions

SaMD is a growing trend within medical device innovation. In this work, we provide the first empirical analysis of SaMD. Within Australia, which relies heavily on importation of medical devices, SaMD shows a greater domestic production than other types of medical devices.

The increased registrations of SaMD has coincided with an increased number of recall events. However, this might be explained by the fact that software can be more readily updated compared to hardware-based devices. Thus, recalls are easier to deal with for SaMD than for a traditional (non-software) medical device. Further research is required to determine the most efficient way that this difference can be reflected in the regulations. With regards to adverse events, it is hard to establish the type of adverse event, as the largest category of events is based on issues originating from the 'Computer Software' itself. This does not help to determine the precise type of event. To this end, adverse event reporting for SaMD might be improved with a greater software-specific granularity in order to better inform regulatory authorities, as well as clinical professionals and patients. More detailed data on this will provide an improved understand of the risks associated with SaMD. It will also allow the regulator to develop further insights into ways to better control these devices.

Author Contributions: Conceptualization, A.C. and J.B.; methodology, A.C. and J.B.; software, A.C.; formal analysis, A.C.; investigation, A.C. and J.B.; resources, A.C.; data curation, A.C. and J.B.; writing—original draft preparation, A.C. and J.B.; writing—review and editing, A.C. and J.B.; visualization, A.C.; project administration, J.B.; funding acquisition, J.B. Both authors have read and agreed to the published version of the manuscript.

Funding: This research was funded by the European Institute of Innovation and Technology (EIT) Health and supported by the National Institute for Health Research (NIHR) Oxford Biomedical Research Centre (BRC).

Data Availability Statement: All data were obtained from public databases and references are provided in the manuscript whenever they are discussed.

Conflicts of Interest: The authors declare no conflict of interest.

References

1. Jamieson, J. Regulation of medical devices involving software in Australia: An overview. In Proceedings of the Sixth Australian Workshop on Safety Critical Systems and Software, Brisbane, Australia, 1 July 2001; Australian Computer Society, Inc.: Darlinghurst, Australia, 2001; Volume 3, pp. 7–12.
2. Therapeutic Goods Administration. Actual and Potential Harm Caused by Medical Software. July 2020. Available online: https://www.tga.gov.au/resource/actual-and-potential-harm-caused-medical-software (accessed on 6 February 2021);
3. IMDRF SaMD Working Group. *Software as a Medical Device (SaMD): Key Definitions*; Tech. Rep.; International Medical Device Regulators Forum; December 2013. Available online: http://www.imdrf.org/docs/imdrf/final/technical/imdrf-tech-131209-samd-key-definitions-140901.pdf (accessed on 6 February 2021).
4. Fu, Z.; Guo, C.; Zhang, Z.; Ren, S.; Jiang, Y.; Sha, L. Study of software-related causes in the fda medical device recalls. In Proceedings of the 2017 22nd IEEE International Conference on Engineering of Complex Computer Systems (ICECCS), Fukuoka, Japan, 6–8 November 2017; pp. 60–69.
5. Ronquillo, J.G.; Zuckerman, D.M. Software-related recalls of health information technology and other medical devices: Implications for fda regulation of digital health. *Milbank Q.* **2017**, *95*, 535–553. [CrossRef] [PubMed]
6. Therapeutic Goods Authority. Therapeutic Goods Authority Website. 2021. Available online: https://www.tga.gov.au/ (accessed on 25 June 2021).
7. International Organization for Standardization. ISO/TS 19218-1:2011Medical Devices—Hierarchical Coding Structure for Adverse Events—Part 1: Event-Type Codes. Available online: https://www.iso.org/standard/55589.html (accessed on 2 April 2021).
8. IEEE Standards Association. IEEE 1044-2009—IEEE Standard Classification for Software Anomalies. Available online: https://standards.ieee.org/standard/1044-2009.html (accessed on 2 April 2021).
9. McHugh, M.; McCaffery, F.; Casey, V. Barriers to adopting agile practices when developing medical device software. In *Software Process Improvement and Capability Determination*; Mas, A., Mesquida, A., Rout, T., O'Connor, R.V., Dorling, A., Eds.; Springer: Berlin/Heidelberg, Germany, 2012; pp. 141–147.
10. Hendricusdottir, R.; Hussain, A.; Milnthorpe, W.; Bergmann, J.H.M. Lack of Support in Medical Device Regulation within Academia. *Prosthesis* **2021**, *3*, 1–8. [CrossRef]
11. Herpin, T.F.; Karuso, H.; Foley, J.E. Australian biotech companies: Navigating the maze. *J. Commer. Biotechnol.* **2005**, *11*, 111–120. [CrossRef]
12. Marucheck, A.; Greis, N.; Mena, C.; Cai, L. Product safety and security in the global supply chain: Issues, challenges and research opportunities. *J. Oper. Manag.* **2011**, *29*, 707–720. [CrossRef]
13. Thirumalai, S.; Sinha, K.K. Product recalls in the medical device industry: An empirical exploration of the sources and financial consequences. *Manag. Sci.* **2011**, *57*, 376–392. [CrossRef]

prosthesis

Review

Regulating Environmental Impact of Medical Devices in the United Kingdom—A Scoping Review

Pranay Arun Kumar

Independent Researcher, New Delhi 110016, India; pranay.arunkumar@gmail.com

Abstract: Medical devices are highly regulated to ensure safety and efficacy of the products and minimize the risk of harm to users and patients. However, the broader impacts of these devices on the environment have scarcely been questioned until recently. The United Kingdom National Health Service intends to achieve a "net zero" emissions service by 2040 and has identified specific targets to achieve through this process. However, medical device manufacturers do not see sufficient incentives to invest in reducing greenhouse gas emissions unless enforced by legislation. Furthermore, there is little evidence on the legislation required to reduce emissions from medical devices. This study addresses the relationship of medical device regulations and the environmental impact of the devices throughout their lifecycle. A scoping review was conducted on academic literature on the topic, followed by a critical review of the current medical device regulations and associated guidelines in the United Kingdom. The challenges to regulating environmental impact of medical devices were identified under seven themes. These challenges were contextualized with the National Health Service target of achieving zero emissions by 2040. The review indicates that current guidelines support single-use disposal of devices and equipment as the best approach to prevent pathogen transmission and landfilling and incineration are the most used waste management strategies. Manufacturers need to be guided and educated on reducing their emissions while ensuring the development of safe and effective devices.

Keywords: medical device; regulations; ecodesign; environmental impact

Citation: Arun Kumar, P. Regulating Environmental Impact of Medical Devices in the United Kingdom—A Scoping Review. *Prosthesis* **2021**, *3*, 370–387. https://doi.org/10.3390/prosthesis3040033

Academic Editor: Jeroen Bergmann

Received: 10 August 2021
Accepted: 1 November 2021
Published: 8 November 2021

1. Introduction

Brexit has ushered in a new era for medical device regulations in the United Kingdom (UK), establishing an independent UK certification system and initiating the phasing out of European certified devices through the Medicines and Medical Devices Act 2021 [1]. As the UK prepares to host the 26th United Nations Climate Change Summit (COP26) [2], there is global attention on how the UK upholds the targets for the reduction of greenhouse gas (GHG) emissions. At the same time, the COVID-19 pandemic has put tremendous pressures on the National Health Service (NHS) to meet the requirements of waves of infections, morbidities and deaths due to the virus and its variants. The UK waste management systems have faced much of the strain due to exponential increases in disposal of medical devices and personal protective equipment (PPE), leading to an increase in the environmental impact of medical equipment [3–5].

Medical devices have significantly contributed to greenhouse gas (GHG) emissions even before the pandemic. In 2019, the GHG emissions from medical equipment procured by the NHS was estimated as 2.52 MtCO_2e, accounting for about 10% of the total emissions from the NHS [6]. Among many reasons for these emissions is the increasing adoption of single-use medical devices, primarily disposed of through incineration or landfilling. The result of this cradle-to-grave lifecycle is not just the emissions generated but also the air, water and soil pollution, damage to biodiversity and contribution to climate change [5,7]. It has been evidenced that climate change has direct implications on human health, and so, it becomes important to mitigate the environmental impacts of this industry to reduce further pressures on healthcare infrastructure.

While medical devices generate GHG emissions throughout their lifecycle, many of these environmental impacts are determined at the early stages of the lifecycle, namely the design and development process [8,9]. There are various barriers to the ecodesign of medical devices, including lack of a regulatory push for ecodesign, high regulatory conformity requirements, lack of knowledge on ecodesign for medical devices and a lack of awareness or education on implementing ecodesign in practice [10–13]. Surveys with designers on this issue indicate that unless there is a regulatory push for environmentally conscious design of medical devices, there is a low probability of ecodesign being considered in the design process [10], but what is not clear is the regulatory push required, how it can be implemented and what are the challenges to enforcing environmentally conscious design of medical devices through regulations. There is little evidence to ascertain how current regulations have affected the environmental impact from medical devices.

This study explores the existing literature on the relationship between medical device regulations and the environmental impact from medical devices in the UK. The study uses academic literature, existing medical device regulations in the UK and the associated guidance provided by the Government of UK to scope the relationship between regulations and the environmental impact of medical devices. The following sections detail the methods used for the scoping review, an assessment of the challenges to regulating environmental impact of this industry and opportunities for further research and ways in which the government can promote the ecodesign of medical devices.

2. Methodology

A literature review was conducted on the state of the art of the role of medical device regulations in environmental implications of medical devices throughout their lifecycles. Keywords included ("environmental legislation" AND "medical device regulations") OR ("medical waste" AND "medical device regulations") OR ("carbon emissions" AND "medical device regulations"). The first 200 results on Google Scholar were considered for each set of keywords. Google Scholar was used as the preferred database due to the wide-ranging subject matter being considered and the scarce data found through previous studies in select databases [11,13]. A further search was conducted on The Web of Science database with the keywords ((ALL = (medical device)) AND ALL = (regulation)) AND ALL = (environmental impact) OR ((ALL = (medical device)) AND ALL = (legislation)) AND ALL = (environmental impact). Fifty-nine results were found. Relevant articles were identified through the title and the contents of the abstract. The inclusion criteria were literature considering environmental impact of medical devices and healthcare infrastructure, medical device regulations and associated legislation. The exclusion criteria were literature discussing safety of healthcare infrastructure not pertaining to environmental impacts, such as regulatory conformity requirements, and pharmacological studies. Papers discussing current practices were also studied to see how regulations and legislation impacted practice.

A second review was conducted on existing regulations for medical devices throughout their lifecycles in the UK. The gov.uk website, which is the UK's public sector information website, was used to conduct a search on the term "medical device". The search was refined, specifying the topic as "health and social care" with a sub-topic of "medicines, medical devices". The content types selected were "guidance and regulation" and "policy papers and consultations". The search resulted in 127 items. The exclusion criteria were documents on legislation for PPE, medicinal products, non-device related care and treatment, medicinal research, medical consultations, competitions, adverse event reports and Northern Ireland related documents. Thirty-eight documents were obtained and studied.

The documents studied helped identify current legislation and regulation on medical devices in the UK, which were validated through the website legislation.gov.uk, the official website for access to all UK legislation. Further searches on gov.uk helped identify guidelines and best practice documents for the various phases of the lifecycle of medical devices. These documents were critically reviewed to identify insights on the current progress in reducing the environmental impact from medical devices and opportunities for further research. In this study, medical devices for which the entire lifecycle of the product is regulated in the United Kingdom were considered. Thus, products that have been manufactured, distributed, purchased, used and disposed of within the UK, not including Northern Ireland, were considered. Devices manufactured or supplied from outside of the UK have not been considered, as different legislations apply to lifecycles beyond those in the United Kingdom. The material extraction stage was not considered in this study because material extraction and synthesis fall under diverse legal acts ranging from the mining of ores, to agriculture, to production of chemicals and polymers, some of which overlap with manufacturing processes and are hence not always bound by any specific legal act. Thus, this process is context dependent and cannot be defined under any specific regulations.

The environmental aspects considered to determine the impact of medical devices on the environment was based on categories provided in the ISO Standard No. 14001:2015, Annex A, including emissions to air, releases to water, waste management, contamination of land, consumption of natural resources and raw materials, chemical releases, toxic substances and other community issues such as noise pollution and release of foul odours [14].

3. Current Regulatory Framework for Medical Devices in the UK

Medical devices are regulated through various regulations throughout their lifecycle in the UK [15–17]. The UK Medical Device Regulations 2002, the main regulatory guidance on medical devices, transposes the European Union (EU) directives on medical devices (Directive 93/42/EEC), active implantable devices (Directive 90/385/EEC) and in-vitro diagnostics (Directive 98/79/EC) regarding the regulatory conformity required to market and sell a device in the UK [18]. These directives should not be confused with the EU regulations for medical devices (EU 2017/745) and in-vitro diagnostic devices (EU 2017/246). While the EU has repealed the directives in favour of new medical device regulations, the UK continues to transpose the earlier European directives, while developing its own regulations and a dedicated UK conformity assessment, independent of European regulatory structures and in line with the Medicines and Medical Devices Act 2021 [1,18–20]. Medical devices must also conform to the General Product Safety Regulations 2005 [21]. While the MDR 2002 regulates manufacture and use of devices, the health technical memoranda (HTM) issued by the Department of Health and Social Care (DHSC) provide specific guidance on the management, decontamination and disposal of devices for healthcare providers, based on a broad range of regulations. The Health Technical Memoranda reviewed in this study include HTM 01-01, 01-05, 01-06 (guidelines on management and decontamination of medical equipment, linen, dental care practice and flexible endoscopes) and 07-01 (guidelines on management of healthcare waste) along with the guidance on reprocessing and re-manufacturing of medical devices [22–38]. For the purpose of this review, it is assumed that the best-practice guidelines meet the requirements of all associated regulations, and reviewing these guidelines along with relevant literature should help scope the challenges based on secondary sources, similar to inferences by Martin et al. for dental care devices [39].

4. Challenges to Regulating Environmental Impact of Medical Devices

Based on the literature reviewed, challenges to regulating environmental impact were identified and structured in seven themes. These challenges were compared with the NHS "net zero" emissions target report and compared with the current research on identified emissions and ways to mitigate these emissions [5]. Specific clauses, their relevance to environmental impact of medical device and associated research opportunities have been provided in Appendix A for further details.

4.1. Psychological Challenges

There are broad psychological explanations to the inertia against environmentally conscious interventions in healthcare. The individual psychological explanations include the paradoxes of preventing harm to the individual while harming the environment, safe management of waste while making the environment unsafe and the use of advanced technology with high energy requirements to treat minor ailments and aesthetic corrections. Topf [40] suggests that the values of profit making and environmentalism are inherently at odds, as profits are driven by consumption. Much of the psychological hindrances in adopting green strategies are also propounded by myths such as greening is costly, a passing fad, not aesthetic or not well supported by the right materials. Then there are forms of denial which limit the greening of hospitals, including a direct denial of contribution to emissions through hospital activities, a procrastination of addressing the problem or resorting to distorting the facts and avoiding relevant information. Beyond the individual psychological explanations to the indifference towards greening hospitals, Topf also suggests group psychological explanations, such as individuals regarding greening as not their responsibility (diffusion of responsibility), disregarding the strides towards greening of other members of a community (pleuritic group ignorance) and the influence of a charismatic leader clouding the individual's judgement, leading to a herd mindset, often resulting in no action (Groupthink) [40].

This need to upskill healthcare workers, and not just sensitize them, has been recognized by the NHS, where 98% of the staff surveyed agree to the need for more sustainable practices in healthcare in the UK. It is well understood that sensitization must be accompanied with education programmes and staff protocol to minimize emissions and maximize value for resources. While various organizations such as the Nursing and Midwifery Council and the General Medical Council have introduced sustainability and climate change into the education system, more evidence is needed for developing protocol for healthcare providers in practice [5].

4.2. Evaluating Emissions and Creating Policy

The NHS quantifies its emissions as per scopes 1, 2 and 3 of the Greenhouse Gas Protocol, based on which they identified medical and non-medical equipment accounting for 18% of the total emission generated by the NHS in 2019 [5,6]. The gold standard for determining environmental impacts of medical devices is to use a Life Cycle Assessment (LCA) [7]. Various competing products can be compared on their impacts, associated costs and value for the healthcare provider to identify the best option. However, the existing literature on comparative assessment of environmental impact of medical devices uses diverse metrics, at various scales of device use and in varied contexts of healthcare settings, state regulations and policy on medical devices. While the quantity of single-use devices used may cause higher emissions than an equivalent reusable version, sterilization processes sometimes are more environmentally damaging as compared to disposing of single-use devices [11,41]. Thus, the results from these LCAs are not easily generalized to suggest policy-level interventions beyond that of the setting in which it has been studied [42–51].

Beyond the concerns of individual, tangible devices, there is limited regulatory oversight for connected health, particularly software as medical devices (SaMD). Data-driven and data-oriented healthcare creates new challenges for the medical device industry, in terms of regulation and the risks involved. The NHS estimates emissions of 456 ktCO_2e from information and communication technology and continued growth in the adoption of digital services [5]. As of now, the understanding of the environmental impacts of SaMD are yet to be determined, even though the role and impact of SaMD in the medical device industry continues to grow [52].

Manufacturers have little incentive in encouraging the regulation of environmental impact of medical devices, unless it is profitable or helps improve the brand image among their consumers [53]. Considering the high conformity requirements to place medical devices on the market, risks of reinfection and cross-contamination and expensive legal action thereafter, manufacturers remain averse to considering environmental impact at the cost of safety and efficacy of devices [10]. This has led to an increasing reliance on single-use disposable medical devices, leading to higher inventory costs at hospitals, higher waste management costs and higher production of medical waste [54,55].

4.3. Lack of Education and Awareness

One factor that has been found to affect the limited involvement of stakeholders on this subject is the lack of awareness and education, both on environmental impact as well as regulatory structures governing medical devices and environmental impact [10]. Kumar and Wang [12] found limited exposure and education of design for environment principles in medical device design and engineering courses around the world. A survey by Moultrie et al. [10] further suggests that designers find current regulations discourage designing for the environment and that manufacturers need to be educated about the opportunities to save expenses in developing sustainable medical devices. Under the theme of knowledge exchange, Martin et al. [39] suggest that there is a lack of encouragement in curricula for sustainable practices in dentistry. Furthermore, while many universities encourage the design and development of medical devices, and provide a platform for research and development in this field, there is very little evidence of education of regulatory structures governing medical devices in the United States of America (USA) and UK academic programmes, as found in a study by Hendricusdottir et al. [56].

4.4. Single-Use, Reusable and Reprocessed Devices

The NHS has found that over 1.4% of all emissions generated are due to single-use devices, some of which can be refurbished and reused to save emissions as well as money. They intend to reduce their reliance on single-use plastics in order to save on waste management costs and almost 224 ktCO_2e in emissions [5]. However, the current regulations suggest otherwise in some cases. Martin et al. note that a significant increase in the generation of biomedical waste in dental practices in the UK over the past few decades can be attributed to increased use of single-use devices and regulation that is confusing staff on best practice and segregation of wastes for sustainable management, among other things [39]. The increasing use of plastics, although providing inexpensive and wide-ranging uses in the medical device industry, is unattractive economically for recycling, thus being relegated to disposal in landfills. The guidelines on decontamination of medical equipment allow healthcare organizations to operate using only single-use devices if they do not have relevant decontamination services [23]. To reduce the risks of prion transmission, certain devices such as endodontic reamers and files which are designated as reusable should be treated as single-use [22,31]. Certain PPE worn during decontamination processes such as aprons, gloves, face masks and gowns must be single-use disposable [22]. Manual cleaning equipment such as brushes and sponges for cleaning endoscopes must be single-use [31]. All accessories with endoscopes must also be single-use. It is also advised that disposable liners be used for decontamination trays [31]. Where the guidelines suggest disposal as the safest form of practice to prevent transmission of pathogens, it is difficult to expect clinicians to identify alternative practices such as treatment or reuse of products.

4.5. Waste Management and the NHS Long Term Plan

The guidelines for safe management of healthcare waste are currently skewed towards landfilling and incineration as the safest options for most of the waste types generated. Table 1 provides an overview of the prescribed disposal strategies for the waste categories provided in HTM 07-01.

Table 1. Disposal options for various waste categories (as prescribed in HTM 07-01 [36]).

Waste Type	Waste Subtypes	Landfill	Municipal Incineration	Energy from Waste	Other Authorised Disposal	Clinical Waste Incineration	Alternative Treatment	Recovery
Domestic type waste		x	x	x	x			x
Offensive waste	Healthcare waste	x	x	x	x			x
	Municipal waste	x	x	x	x			x
Anatomical waste	Chemically preserved					x		
	Not chemically preserved					x		
Infectious waste	Contaminated with chemicals					x		
	Not containing contaminated chemicals or medicinal contamination					x	x	
Sharps	Non-medicinally contaminated					x	x	
	Medicinally contaminated other than cytotoic and cytostatic waste					x		
	Contaminated with cytotoxic and cytostatic waste					x		
Other infectious waste contaminated with cytotoxic and cytostatic waste						x		
Cytotoxic and cytostatic medicines	(in original packaging)					x		
	(not in original packaging)					x		
Other medicines	(in original packaging)					x		
	(not in original packaging)					x		
Dental amalgam								x

Table 1. *Cont.*

Waste Type	Waste Subtypes	Landfill	Municipal Incineration	Energy from Waste	Other Authorised Disposal	Clinical Waste Incineration	Alternative Treatment	Recovery
Photographic (X-ray) waste	X-ray fixer						x	x
	X-ray developer						x	x
	Lead foil							x
	X-ray film							x
Gypsum and plaster-cast waste		x (specialist landfill)						x
Radioactive waste						x		

Out of the 22 waste types, 16 are advised for incineration (in red) as one of the strategies for waste management. 4 of the 22 waste types are advised for landfilling (in yellow). Thus, a combined total of 20 of the 22 waste types are advised for a cradle-to-grave lifecycle (either incineration or landfilling). Nine out of the 22 waste types can be potentially recovered in some form (in green). Out of the nine, only three waste types are necessary to be recovered (dental amalgam, lead foil and X-ray film). The other three may still be disposed of without any recovery strategy. Thus, there is a predominant leaning towards cradle-to-grave lifecycles with few recovery options prescribed for various waste categories.

Healthcare waste management also faces new challenges which have currently not been addressed by the prescribed guidelines. The NHS has embarked on a long-term redesign of the care pathways it offers to the UK, specifically towards reducing in-person visits for patients through digital care consultations and reducing the burden on critical care infrastructure through preventive and public health investments [57]. This also indicates the increasing reliance on home healthcare. However, the current guidelines do not address the appropriate management of hazardous waste generated through home healthcare. Waste that contains hazardous substances such as cytotoxic or cytostatic medication or offensive waste that is infectious in nature is deemed as healthcare waste [36]. However, when these substances are disposed of through municipal waste streams, they are treated as municipal waste. There is also a broader acceptance and uptake of implants by society, both functional and aesthetic. However, these implants are not treated as healthcare waste, unless they are identified through a post-mortem or registered for donation upon the death of the current user [36]. The cremation or burial of these implants poses further environmental challenges which have not been studied or addressed within the current waste management policies.

4.6. Lack of Environmentally Conscious Standards for Medical Device Design

Another factor that has to be considered is the designated standards for medical devices, in-vitro diagnostic devices and active implantable devices. The DHSC does not include the ISO 14000 series on environmental management [58] or the IEC 62430 on environmentally conscious design (ECD) [59] as designated standards, thus excluding any standards on environmental impact of medical devices [60–62]. ISO 14006 builds on the existing quality management system of an organization (ISO 9001) [63] and while the UK designates a standard for quality management for medical devices (EN ISO 13485:2016) it does not mandate adherence to ecodesign or ECD standards. In fact, none of the standards for ecodesign endorsed or prescribed by the British standards institution have been designated for medical devices by the DHSC [64] nor has the ecodesign directive (The

Ecodesign for Energy-Related Products Regulations 2010) [65] been referenced in the MDR 2002 [18,66].

4.7. Limitations to Legislation Motivating Environmentally Conscious Practices

Assuming the above-mentioned challenges can be addressed, there is still the question of whether regulations and legislation can reduce the environmental impact of medical devices. Martin et al. identified that in the UK, the implementation of HTM 01-05 [22] led to an increase in waste management costs due to the instruction of disposing PPEs and single-use devices to prevent reinfection and cross-infection cases. The interpretation of HTM 01-05 has also frustrated users, leading to confusion in sustainable practices while trying to avoid litigation. Martin et al. [39] find that the legislation and regulation of safe disposal of dental amalgam is varied and inconsistent across the world, leading to the risk of higher contribution of mercury toxicity in the environment. Technology, in the form of dental separators, exists to ensure safe disposal of mercury, but this technology is not used all over the world, and there continues to be resistance from the dental profession to mandate dental separators in legislation. Along with metals, X-ray waste, gypsum and composite waste, there is a lack of harmonized regulations for the safe management and disposal of these materials from the dental industry. Wagner [67] argues that environmental legislation incentivizes actors to conceal relevant information of the harm that their products may cause to the environment, despite government subsidies on research on factors affecting environmental damage. Wagner also proposes that regulatory bodies cannot wait for the research to emerge regarding harmful substances (as has been the modus operandi), and instead they must penalize the concealment of information regarding the harmful impacts of substances being used or produced by various organizations. It has become evident over the last two decades that the pace of scientific progress has been accelerated in comparison with the legislation to control its adverse effects. While scientific progress cannot be slowed down, new approaches are required to increase the pace of legislation to ensure safe and effective use of new technology [68]. Musazzi et al. [69] argue that the current European regulatory framework does not effectively assess the human health and environmental risks of nanomaterials in medical devices, thus posing risks to users and patients. Ren et al. [70] studied the effect of environmental regulations on eco-efficiency gains in different regions of China. They classified regulations under three categories: command-and-control regulations (legislation discouraging environmentally damaging practices), market-based regulations (incentivizing eco-efficiency through tax-rebates, taxes and emission subsidies) and voluntary regulation (guidelines and protocols that are encouraged through public participation but not imposed as legislation). Their study indicates that different types of regulations have shown to influence eco-efficiency differently in different regions of China. Based on their study, they were able to propose suitable policy interventions to specific regions of China based on public participation, role of incentives and role of regulatory discouragement of environmentally damaging practices. The current research indicates that there is no consensus on how climate change policies can be implemented to reduce the environmental impact from healthcare systems. Yet, the NHS continues to be one of the few healthcare systems in the world with a meticulous record of environmental impacts and a long term plan for achieving net zero emissions through its services [5,6].

The review of literature on the relationship between medical device regulations and environmental impact, although insightful in terms of the challenges to regulating environmental impact, provided little evidence of how environmental legislation works to curb environmental impact and how such legislation can be developed. Fragmented literature indicated that research is required to structure the process and methods for translating knowledge of environmental implications of medical devices to policy for more transformative change.

5. Opportunities for Future Research and Policy Development

This review has helped identify various gaps in the current literature for regulating environmental impact in the medical device industry. The results indicate that more research is required to understand how the environmental impact of medical devices can be regulated without compromising the safety and efficacy of the devices. Through the findings, a few suggestions for research directions have been proposed which may encourage medical device stakeholders to embrace environmentally conscious approaches to their trade.

While many studies exist on the impact assessment of individual medical devices within their defined contexts, it is not clear how the evolution of the regulatory framework affects the environmental impacts of medical devices. It is also not clear how this can be studied. Yet it is important to develop systematic assessment methods so that future regulations can be developed with clear evidence of associated environmental impact. The NHS expects an increase in emissions of 1734 $ktCO_2e$ from vehicle use [5]. However, some of these emissions can be reduced through supply chain initiatives [5]. The adoption of industry 4.0 strategies provides pathways to reduce transport requirements by encouraging in-house manufacture and reprocessing of devices [13]. As the current regulations allow both of these processes, research can help develop complete cradle-to-cradle device lifecycles and product-service systems within healthcare institutions, reducing transport emissions of supply chain requirements. The MDR 2002 also provides conformity requirements for in-house manufacture of medical devices by healthcare facilities, for which the overall conformity requirements are lower than those for externally manufactured devices [18,71]. This also contributes to the envisioned reduction of emissions by the NHS from metal instrument reprocessing (157 $ktCO_2e$) and device reuse and refurbishment (202 $ktCO_2e$) [5].

The NHS intends to reduce reliance on single-use plastics and increase reuse and refurbishment of medical devices [5]. However, these targets are dependent on appropriate procurement and supply chain transformations. Currently, these impacts are being assessed through a limited number of suppliers volunteering to share their plans on carbon reduction [5], but more policy level decisions are required to achieve the goal of net zero emissions within the stipulated timeline. From a manufacturer's perspective, there are three approaches to encourage sustainability in the medical device ecosystem. The first is to encourage the recycling of materials after the use of a product as well as increase the use of recycled materials for the production of new devices. The second is to adopt more sustainable practices in manufacturing, such as reducing the waste of material in production processes and use local production facilities and local supply chains. The third approach is to design sustainability into the entire lifecycle of the product. This would enable considering the materials used, the logistics involved in cradle-to-cradle design and ensure that the product is easy to disassemble [11,53]. As the government of the UK already provides guidance on conforming to the various regulations, and also on the design of medical devices [72], guidance can be developed for manufacturers on environmentally conscious design of medical devices, based on established standards as well as through research by field experts [59,63]. The government also provides guidance for patients and users on the use and management of medical devices [73], particularly for home use and devices prescribed by clinicians. Thus, users can also be educated on the environmental impacts of medical devices through reports, documents and leaflets at the local healthcare facility, generating an informed demand for more environmentally conscious practices in the design of medical devices.

Through the investments in a digital care pathway redesign, the NHS expects to reduce travel-related emissions by 159 $ktCO_2e$ [5]. However, the NHS also estimates emissions of 456 $ktCO_2e$ from information and communication technology [5]. Currently there is very little research on assessing the environmental impact of SaMD. Considering the rising dependence on digital health records, and information engineering approaches to healthcare, it will be important to identify critical factors influencing the environmental impacts of SaMD and address them through regulatory oversight, particularly if the NHS strives to build a net zero digital maturity framework [5].

6. Conclusions

The increasing global focus on climate change begs the question of environmental sustainability of healthcare systems. The growing evidence of rising GHG emissions from the medical device industry, and the paradox of harm to the individual versus harm to the environment of the population questions which strategies will tackle the inertia against ecodesign of medical devices. The existing research suggests that regulating environmental impacts of medical devices is necessary for compliance of the industry as a whole, but the evidence on how to regulate environmental impacts is limited.

This study identified the various challenges to regulating environmental impact of medical devices and how the current regulations can affect environmental sustainability of this industry. Lower emissions cannot be promoted at the cost of safety and efficacy of devices, and manufacturers will not voluntarily consider environmental sustainability at the cost of economic profit. There is also limited evidence to suggest that regulation will help limit environmental impacts, and it may lead to manufacturers hiding the dangerous environmental impacts of their trade. While it is clear that the industry has significant environmental impacts and the NHS strives to work towards becoming a net-zero emissions organization, the appropriate strategies for the medical device industry are continuing to evolve.

Current guidelines on management and decontamination of devices actively promote disposal of devices after a single use to prevent reinfection and cross-contamination, particularly from prion transmission. The regulations on disposal of devices indicate incineration and landfilling as the best practice, with very few opportunities for waste recovery. Despite the existence of recognized ecodesign and environmentally conscious design standards, these have not yet been designated to medical device regulations.

By educating stakeholders such as manufacturers and patients, more awareness can be generated on environmentally conscious approaches to the management of medical devices throughout their lifecycle. The government can support ecodesign of devices through appropriate guidance and eventually regulate the impacts based on developed criteria for evaluation of environmental impacts.

The regulation of environmental impacts of medical devices is a complex issue, with many factors working at cross-purposes with each other. More research is required to understand how this industry can accommodate environmentally conscious practices which are safe, effective and economically sustainable. While this is an opportune moment for the UK to consider ways to improve their medical device regulations, the literature reviewed indicates a lack of attention to environmental impact of medical devices around the world and a lack of appropriate legislation to curb these impacts. Climate change and healthcare are global challenges with global implications, as has been seen through the COVID-19 pandemic and the impact of climate change on destructive weather patterns. These challenges are also closely interlinked, suggesting that curbing environmental impacts from the medical device industry can go a long way in ensuring sustainability of healthcare systems.

Funding: This research received no external funding.

Acknowledgments: I would like to thank the reviewers and the academic editor for their rich feedback in the development of this paper. I would also like to thank the managing editor for supporting the submission and revision processes.

Conflicts of Interest: The author declares no conflict of interest.

Appendix A

Table A1. Critical review of regulatory clauses and research opportunities.

Legal Act	Statement	Relevance to Environmental Impact	Research Opportunities
HTM 01-01, Part-A (pg 10)	"Section 3 Guidance for commissioners, regulators and providers, point 3.2: Responsibility for achieving acceptable standards of decontamination rests with commissioning organisations, individual trusts and provider organisations. Reprocessing units in healthcare establishments responsible for the decontamination of medical devices fall into two distinct categories when considering compliance with the MDD: • Devices transferred between legal entities (for example—reprocessing by one entity followed by use in another). • Devices remaining within one legal entity (for example—reprocessing and use by the same entity or organisation)." [23]	Reprocessing medical devices has been evidenced to incur reduced environmental impacts as compared with the equivalent disposable options for certain devices [48,74]. However, research does not indicate the environmental implications of on-site and off-site reprocessing.	Research on the environmental and cost implications of on-site and off-site reprocessing of medical devices can help healthcare centres consider investment strategies in reprocessing of devices.
HTM 01-01, Part-A (pg 15)	"Section 4 Regulatory framework, Outsourcing 4.23 The options for those healthcare organisations that do not undertake decontamination services include: • Using a decontamination service that is registered with the MHRA, that is compliant with the MDR, and that uses a notified body as its third-party auditor. • Using CE-marked single-use medical devices." [23]	There are varied reports on the environmental and cost implications of reusable vs. single-use devices [49,75]. This clause of the memorandum indicates that health systems are allowed to run completely on single-use devices if no decontamination facility is available. There is limited literature to indicate the environmental and cost implications of health systems of a similar scale running on purely single-use devices or having access to decontamination/ reprocessing facilities.	Evaluating the environmental and cost implications of running a healthcare facility purely on single-use devices as compared to investing in decontamination and reprocessing systems.
HTM 01-01, Part-A (pg 22)	"6 Management of surgical instruments, Loan sets 6.11 Instrument sets that are supplied from an external source, used for that procedure only and then returned are known as loan sets. This practice increases the risks associated with the decontamination and reprocessing of such instruments, because the organisation may not be familiar with them. Organisations have also expressed concern over the decontamination status of such instruments and the lack of track and traceability, including potential for instrument migration. It is a requirement of the Code of Practice that reusable medical devices should be decontaminated in accordance with manufacturers' instructions. Therefore, loan sets should be provided with decontamination instructions so that staff can ensure their compatibility with local decontamination processes. It should be ensured that when equipment is supplied to a healthcare provider, adequate time is allowed for cleaning, sterilization and return of the equipment to the theatres, both prior to and after use (see the AfPP's (2010) guidance 'Loan set management principles between suppliers/manufacturers, theatres & sterile service departments' and MHRA's 'Managing medical devices')." [23]	Loaning of medical devices allows sharing of resources, reducing the reliance on procuring new devices for each healthcare setting. It is well established that a sharing economy promotes sustainable outcomes and reduces environmental impacts in various industries such as mobility, digital economies and consumer appliances [76]. However, studies do not indicate the environmental and cost impacts of a sharing economy in healthcare, particularly the case of loaning medical devices vs. procuring devices.	Evaluating the environmental and cost implications of loaning medical devices vs. procuring medical devices for the same purpose.

Table A1. *Cont.*

Legal Act	Statement	Relevance to Environmental Impact	Research Opportunities
HTM 01-05 (pg 14)	"2 Essential quality requirements and best practice, Segregating instruments 2.17 Where instruments are difficult to clean, consideration should be given to replacing them with single-use instruments where possible. In dentistry this will include, but is not limited to, instruments such as matrix bands, saliva ejectors, aspirator tips and three-in-one tips. 2.18 Where endodontic reamers and files are designated reusable, they should be treated as single patient use or single use—regardless of the manufacturer's designation—to reduce the risk of prion transmission. Practices must have effective procedures in place to exclude errors in identifying the instrument(s) and associating them with the correct patient." [22]	When reusable devices are replaced with single-use devices due to difficulties in cleaning them, it is a design failure leading to the adoption of more wasteful alternatives. However, research does not indicate these design failures and the resultant transition to single-use devices.	Identify devices that are difficult to decontaminate effectively and study the design failures leading to a replacement with single-use devices.
HTM 01-05 (pg 34)	"6 General hygiene principles, Personal protective equipment for decontamination processes 6.14 Appropriate PPE should be worn during decontamination procedures. PPE includes disposable clinical gloves, household gloves, plastic disposable aprons, face masks, eye protection and adequate footwear. PPE should be stored in accordance with manufacturers' instructions. 6.21 Gloves other than domestic household types are single use only. They should be discarded as clinical waste. 6.25 Aprons should be used as a single-use item and disposed of as clinical waste. Plastic aprons should be changed at the completion of each procedure. 6.27 Face masks are single-use items and should be disposed of as clinical waste. 6.29 Eye protection may be reusable but is often difficult to clean. It may be reused if cleaned according to manufacturers' instructions. This should take place when it becomes visibly dirty and/or at the end of each session. Disposable visors are available and may be used. 6.33 Short sleeves allow the forearms to be washed as part of the hand hygiene routine. Dental staff need to be aware of the hazards that may be encountered in the decontamination process and may wish to wear long-cuffed gloves or disposable long-sleeved gowns to protect their arms." [22]	Personal protective equipment has been a major cause of excess waste and environmental impact through the COVID-19 pandemic, and the current guidance also endorses disposal of various PPE after a single use. However, the guidance does not necessitate the disposal of all PPE, and there is no argument provided for single use or reuse of equipment.	Research is needed to evaluate the risks of cross-contamination from various PPE, and appropriate design criteria is required to ensure that equipment is designed appropriately for minimum waste.

Table A1. *Cont.*

Legal Act	Statement	Relevance to Environmental Impact	Research Opportunities
HTM 01-06 (pg 3)	"2 Flexible endoscopes and decontamination, 2.7 The process of decontaminating flexible endoscopes with lumens has three components: a. Manual cleaning: this includes brushing with a specific single-use cleaning device, rinsing and exposure of all external and accessible internal components to a lowfoaming detergent known to be compatible with the endoscope. This procedure is uncontrolled and relies on the training of the operator for success." [31]	Point 2.7a specifies the use of a single-use cleaning brush; however, research does not indicate the associated value over reusable brushes. Furthermore, research does not indicate environmental or cost advantages over reusable brushes.	Compare the environmental and cost implications of single-use versus reusable channel port cleaning brushes and determine the risk versus benefits of the two.
HTM 01-06 (pg 12)	"5 Human prion diseases (including variant CJD and other forms of CJD) 5.17 The guidance below is based on that from the ACDP-TSE Subgroup's Annex F (last revised in October 2015). Users should check for updates on the ACDP-TSE Subgroup's website. a. Channel cleaning brushes and, if biopsy forceps or other accessories have been passed, the valve on the endoscope biopsy/instrument channel port should be disposed of as healthcare waste after each use. Single-use biopsy forceps should be used in all patients. Endoscope accessories should be single use wherever possible. It is essential to have systems in place that enable endoscopes, together with all their detachable components and any reused accessories, to be traced to the patients on whom they have been used. f. Following use in patients at risk of vCJD endoscopic accessories (including normally reusable devices such as heater probes) and cleaning aids such as brushes should be disposed of as healthcare waste." [31]	To reduce the risk of prion transmission, this point indicates disposal of cleaning equipment, which would either lead to landfilling or incineration. However, literature does not indicate whether alternative options supporting a cradle-to-cradle lifecycle exist for these products.	To explore alternative recovery and treatment strategies for disposable cleaning equipment for medical devices
HTM 01-06, Part B (pg 4)	"1 Design of an endoscope reprocessing unit, Layout of the unit, Single-room decontamination area 1.13 In addition to endoscope decontamination, the decontamination of trays or use of disposable liners is recommended. In addition, transport trolleys should be considered for decontamination as necessary. This should be considered as part of operational risk assessment." [31]	Packaging plastics are extensively used in healthcare settings for pre-sterilized as well as non-sterile products to ensure safe handling of equipment by healthcare workers and reduce the risks of cross-contamination. However, the environmental impact of these liners has not been evidenced, considering most of the packaging is disposed of after a single use. It has been evidenced that packaging in other industries is one of the leading producers of landfill waste.	Identify, develop and comparatively evaluate suitable alternatives to disposable liners for medical devices

Table A1. *Cont.*

Legal Act	Statement	Relevance to Environmental Impact	Research Opportunities
HTM 07-01 (pg 23)	4 Healthcare waste definitions and classifications Healthcare waste classification and assessment framework [36]	Waste having medicinal properties (e.g., expired medicines, devices containing medicinal products) produced from households is treated as municipal waste, despite being assessed by the guidelines as healthcare waste. The environmental impacts of home healthcare waste have scarcely been studied, despite having risks of leaching hazardous substances into municipal landfills, soil, air and water tables. The safe management of home healthcare waste has not been addressed in these guidelines.	Evaluate the quantities and environmental impacts of home healthcare waste.
HTM 07-01 (pg 46)	"Implants 4.154 Special care should be taken when removing an implant, particularly if it has electronic components such as an implantable cardioverter defibrillator or other implanted cardiac aid. For example: • there may be a risk of electric shock to a person removing and subsequently handling them; • cremation or disposal by incineration might cause batteries to explode, leaking toxic gas. 4.155 Such implants should be deactivated, removed with consent, decontaminated, and disposed of in a safe manner in the hazardous waste stream. Note Removed items are waste produced by the healthcare organisation. Where the patient has asked to retain the item, it is not considered waste, since it has not been discarded. 4.156 Protocols for the removal of implants should be determined locally. Local cardiac units, manufacturers/suppliers and funeral directors should be consulted. Helpful guidance has been published by the Association of British Healthcare Industries, the National Association of Funeral Directors, the Institute of Cemetery and Crematorium Management, and the Medicines and Healthcare products Regulatory Agency (MHRA) in its circular MDA SN 2008/068). 4.157 Disposal may include return to the manufacturer or cardiac unit to access stored data (see also Chapter 5, 'Waste minimisation, segregation, colour-coding and storage'). The receiving authority needs to be aware of duty-of-care implications. Reference to decontamination procedures and appropriate protocols for returning equipment should be provided by the receiving authority." [36]	Current research on implants from deceased persons only relates to organ donation, person identification through implants and material recovery from post-mortems [77–80]. However, research does not address the environmental impact of implants which are not safely disposed of by healthcare facilities. The increasing access and affordability of implants, both functional and aesthetic, make it an important aspect of study from an environmental impact perspective.	Evaluating the environmental impact of body implants throughout their lifecycle and developing suitable recovery strategies and device designs to reduce associated waste.

Table A1. *Cont.*

Legal Act	Statement	Relevance to Environmental Impact	Research Opportunities
HTM 07-01 (pg 53–57)	"Figure 11. Waste segregation chart" [36]	The waste segregation chart not only delineates the segregation process but also provides the disposal options. The disposal options for each waste type have been provided in Table 1. Majority of the disposal strategies suggest cradle-to-grave lifecycles, with very few recovery strategies offered for different types of waste streams.	Explore novel waste recovery and value addition strategies for waste types currently designated for a cradle-to-grave lifecycle.

References

1. Medicines and Medical Devices Act. 2021. Available online: https://www.legislation.gov.uk/ukpga/2021/3/contents/enacted (accessed on 20 September 2021).
2. UN Climate Change Conference. 2021. Available online: https://ukcop26.org/ (accessed on 20 September 2021).
3. You, S.; Sonne, C.; Ok, Y.S. COVID-19's Unsustainable Waste Management. *Science* **2020**, *368*, 1438. [CrossRef] [PubMed]
4. Sarkodie, S.A.; Owusu, P.A. Global Assessment of Environment, Health and Economic Impact of the Novel Coronavirus (COVID-19). *Environ. Dev. Sustain.* **2021**, *23*, 5005–5015. [CrossRef] [PubMed]
5. National Health Service. *Delivering a 'Net Zero' National Health Service*; National Health Service: London, UK, 2021; p. 75.
6. Tennison, I.; Roschnik, S.; Ashby, B.; Boyd, R.; Hamilton, I.; Oreszczyn, T.; Owen, A.; Romanello, M.; Ruyssevelt, P.; Sherman, J.D.; et al. Health Care's Response to Climate Change: A Carbon Footprint Assessment of the NHS in England. *Lancet Planet. Health* **2021**, *5*, e84–e92. [CrossRef]
7. MacNeill, A.J.; Hopf, H.; Khanuja, A.; Alizamir, S.; Bilec, M.; Eckelman, M.J.; Hernandez, L.; McGain, F.; Simonsen, K.; Thiel, C.; et al. Transforming the Medical Device Industry: Road Map to A Circular Economy: Study Examines a Medical Device Industry Transformation. *Health Aff.* **2020**, *39*, 2088–2097. [CrossRef] [PubMed]
8. Ramani, K.; Ramanujan, D.; Bernstein, W.Z.; Zhao, F.; Sutherland, J.; Handwerker, C.; Choi, J.-K.; Kim, H.; Thurston, D. Integrated Sustainable Life Cycle Design: A Review. *J. Mech. Des.* **2010**, *132*, 091004. [CrossRef]
9. Jeswiet, J.; Hauschild, M. EcoDesign and Future Environmental Impacts. *Mater. Des.* **2005**, *26*, 629–634. [CrossRef]
10. Moultrie, J.; Sutcliffe, L.; Maier, A. Exploratory Study of the State of Environmentally Conscious Design in the Medical Device Industry. *J. Clean. Prod.* **2015**, *108*, 363–376. [CrossRef]
11. Arun Kumar, P. EcoDesign for Medical Devices: Barriers and Opportunities to Eco Effective Design of Medical Devices. Master's Thesis, Royal College of Art, London, UK, 2020.
12. Kumar, P.A.; Wang, S.J. Encouraging DfE in Design Education to Promote Sustainable Medical Product Design. In *Designing Sustainability for All*; Edizioni POLI.Design: Milan, Italy, 2019; Volume 4, pp. 1354–1359.
13. Arun Kumar, P.; Wang, S.J. The Design Intervention Opportunities to Reduce Procedural-Caused Healthcare Waste Under the Industry 4.0 Context—A Scoping Review. In *Interactivity and Game Creation*; Brooks, A., Brooks, E.I., Jonathan, D., Eds.; Lecture Notes of the Institute for Computer Sciences, Social Informatics and Telecommunications Engineering; Springer International Publishing: Cham, Switzerland, 2021; Volume 367, pp. 446–460. [CrossRef]
14. International Organization for Standardization. Environmental Management Systems-Requirements with Guidance for Use (ISO Standard No. *14001:2015)*. Available online: https://www.iso.org/standard/60857.html (accessed on 20 September 2021).
15. Medicines & Healthcare Products Regulatory Agency. Managing Medical Devices Guidance for Health and Social Care Organisations. Available online: https://www.gov.uk/government/publications/managing-medical-devices (accessed on 20 September 2021).
16. Medicines & Healthcare Products Regulatory Agency. Management of In Vitro Diagnostic Medical Devices. Available online: https://www.gov.uk/government/publications/in-vitro-diagnostic-medical-devices-procurement-safety-quality-and-performance (accessed on 20 September 2021).
17. Medicines & Healthcare Products Regulatory Agency. Management and Use of IVD Point of Care Test Devices. Available online: https://www.gov.uk/government/publications/in-vitro-diagnostic-point-of-care-test-devices/management-and-use-of-ivd-point-of-care-test-devices (accessed on 20 September 2021).
18. The Medical Device Regulations. 2002. Available online: https://www.legislation.gov.uk/uksi/2002/618/contents (accessed on 20 September 2021).

19. European Parliament. Regulation (EU) 2017/745 of the European Parliament and of the Council of 5 April 2017 on Medical Devices, Amending Directive 2001/83/EC, Regulation (EC) No 178/2002 and Regulation (EC) No 1223/2009 and Repealing Council Directives 90/385/EEC and 93/42/EEC. 2017, pp. 1–175. Available online: https://eur-lex.europa.eu/legal-content/EN/TXT/PDF/?uri=CELEX:32017R0745 (accessed on 20 September 2021).
20. European Parliament; Council of the European Union. Regulation (EU) 2017/746 of the European Parliament and of the Council. 2017. Available online: https://www.legislation.gov.uk/eur/2017/746/contents (accessed on 20 September 2021).
21. Medicines & Healthcare Products Regulatory Agency. Medical Devices: The Regulations and How We Enforce Them. Available online: https://www.gov.uk/government/publications/report-a-non-compliant-medical-device-enforcement-process/how-mhra-ensures-the-safety-and-quality-of-medical-devices (accessed on 20 September 2021).
22. Department of Health. *Health Technical Memorandum 01-05: Decontamination in Primary Care Dental Practices*; Government of the United Kingdom: London, UK, 2013.
23. Department of Health. *Health Technical Memorandum 01-01: Management and Decontamination of Surgical Instruments (Medical Devices) Used in Acute Care Part A: Management and Provision*; Government of the United Kingdom: London, UK, 2016.
24. Department of Health. *Health Technical Memorandum 01-01: Management and Decontamination of Surgical Instruments (Medical Devices) Used in Acute Care Part B: Common Elements*; Government of the United Kingdom: London, UK, 2016.
25. Department of Health. *Health Technical Memorandum 01-01: Management and Decontamination of Surgical Instruments (Medical Devices) Used in Acute Care Part C: Steam Sterilization*; Government of the United Kingdom: London, UK, 2016.
26. Department of Health. *Health Technical Memorandum 01-01: Management and Decontamination of Surgical Instruments (Medical Devices) Used in Acute Care Part E: Alternatives to Steam for the Sterilization of Reusable Medical Devices*; Government of the United Kingdom: London, UK, 2016.
27. Department of Health. *Health Technical Memorandum 01-04: Decontamination of Linen for Health and Social Care—Management and Provision*; Government of the United Kingdom: London, UK, 2016.
28. Department of Health. *Health Technical Memorandum 01-04: Decontamination of Linen for Health and Social Care—Engineering, Equipment and Validation*; Government of the United Kingdom: London, UK, 2016.
29. Department of Health. *Health Technical Memorandum 01-04: Decontamination of Linen for Health and Social Care. Social Care*; Government of the United Kingdom: London, UK, 2016.
30. Department of Health. *Health Technical Memorandum 01-04: Decontamination of Linen for Health and Social Care. Guidance for Linen Processors Implementing BS EN 14065*; Government of the United Kingdom: London, UK, 2016.
31. Department of Health. *Health Technical Memorandum 01-06: Decontamination of Flexible Endoscopes. Part A: Policy and Management*; Government of the United Kingdom: London, UK, 2016.
32. Department of Health. *Health Technical Memorandum 01-06: Decontamination of Flexible Endoscopes. Part B: Design and Installation*; Government of the United Kingdom: London, UK, 2016.
33. Department of Health. *Health Technical Memorandum 01-06: Decontamination of Flexible Endoscopes. Part C: Operational Management*; Government of the United Kingdom: London, UK, 2016.
34. Department of Health. *Health Technical Memorandum 01-06: Decontamination of Flexible Endoscopes. Part D: Validation and Verification (Including Storage/Drying Cabinets)*; Government of the United Kingdom: London, UK, 2016.
35. Department of Health. *Health Technical Memorandum 01-06: Decontamination of Flexible Endoscopes. Part E: Testing Methods*; Government of the United Kingdom: London, UK, 2016.
36. Department of Health. *Environment and Sustainability Health Technical Memorandum 07-01: Safe Management of Healthcare Waste*; Stationery Office: London, UK, 2006.
37. Medicines & Healthcare products Regulatory Agency. *Single-Use Medical Devices: UK Guidance on Re-Manufacturing*; Medicines and Healthcare products Regulatory Agency: London, UK, 2016.
38. Medicines & Healthcare products Regulatory Agency. *Single-Use Medical Devices: Implications and Consequences of Reuse*; Medicines and Healthcare products Regulatory Agency: London, UK, 2019.
39. Martin, N.; Sheppard, M.; Gorasia, G.; Arora, P.; Cooper, M.; Mulligan, S. Awareness and Barriers to Sustainable Dental Practice—A Scoping Literature Review. *J. Dent.* **2021**, 103735. [CrossRef]
40. Topf, M. Psychological Explanations and Interventions for Indifference to Greening Hospitals. *Health Care Manag. Rev.* **2005**, *30*, 2–8. [CrossRef]
41. I'ons, G. From Design to Disposal: Achieving Sustainability in Medical Devices. *On Drug Delivery.* **2020**, *112*, 62–65. Available online: https://www.ondrugdelivery.com/from-design-to-disposal-achieving-sustainability-in-medical-devices/ (accessed on 20 September 2021).
42. Dettenkofer, M.; Grießhammer, R.; Scherrer, M.; Daschner, F. Einweg- versus Mehrweg-Patientenabdeckung im Operationssaal Ökobilanz: Vergleich von Zellstoff-Polyethylen- und Baumwoll-Mischabdeckung: Ökobilanz: Vergleich von Zellstoff-Polyethylen- und Baumwoll-Mischabdeckung. *Chirurg* **1999**, *70*, 485–491. [CrossRef] [PubMed]
43. Ison, E.; Miller, A. The use of lca to introduce life-cycle thinking into decision-making for the purchase of medical devices in the NHS. *J. Environ. Assess. Policy Manag.* **2000**, *2*, 453–476. [CrossRef]
44. Stripple, H.; Westman, R.; Holm, D. Development and Environmental Improvements of Plastics for Hydrophilic Catheters in Medical Care: An Environmental Evaluation. *J. Clean. Prod.* **2008**, *16*, 1764–1776. [CrossRef]

45. McGain, F.; McAlister, S.; McGavin, A.; Story, D. The Financial and Environmental Costs of Reusable and Single-Use Plastic Anaesthetic Drug Trays. *Anaesth. Intensive Care* **2010**, *38*, 538–544. [CrossRef] [PubMed]
46. Unger, S.R.; Landis, A.E. Comparative Life Cycle Assessment of Reused versus Disposable Dental Burs. *Int. J. Life Cycle Assess.* **2014**, *19*, 1623–1631. [CrossRef]
47. Thiel, C.L.; Eckelman, M.; Guido, R.; Huddleston, M.; Landis, A.E.; Sherman, J.; Shrake, S.O.; Copley-Woods, N.; Bilec, M.M. Environmental Impacts of Surgical Procedures: Life Cycle Assessment of Hysterectomy in the United States. *Environ. Sci. Technol.* **2015**, *49*, 1779–1786. [CrossRef] [PubMed]
48. Unger, S.; Landis, A. Assessing the Environmental, Human Health, and Economic Impacts of Reprocessed Medical Devices in a Phoenix Hospital's Supply Chain. *J. Clean. Prod.* **2016**, *112*, 1995–2003. [CrossRef]
49. McGain, F.; Story, D.; Lim, T.; McAlister, S. Financial and Environmental Costs of Reusable and Single-Use Anaesthetic Equipment. *Br. J. Anaesth.* **2017**, *118*, 862–869. [CrossRef] [PubMed]
50. Unger, S.R.; Hottle, T.A.; Hobbs, S.R.; Thiel, C.L.; Campion, N.; Bilec, M.M.; Landis, A.E. Do Single-Use Medical Devices Containing Biopolymers Reduce the Environmental Impacts of Surgical Procedures Compared with Their Plastic Equivalents? *J. Health Serv. Res. Policy* **2017**, *22*, 218–225. [CrossRef] [PubMed]
51. Sherman, J.D.; Raibley, L.A.; Eckelman, M.J. Life Cycle Assessment and Costing Methods for Device Procurement: Comparing Reusable and Single-Use Disposable Laryngoscopes. *Anesth. Analg.* **2018**, *127*, 434–443. [CrossRef] [PubMed]
52. Carroll, N.; Richardson, I. Software-as-a-Medical Device: Demystifying Connected Health Regulations. *J. Syst. Inf. Technol.* **2016**, *18*, 186–215. [CrossRef]
53. I'ons, G. Taking the Long View: Harnessing Product Lifecycle Drives Sustainability in Medical Devices. *MedTech Intelligence*. 23 April 2021. Available online: https://www.medtechintelligence.com/column/taking-the-long-view-harnessing-product-lifecycle-drives-sustainability-in-medical-devices/ (accessed on 20 September 2021).
54. Zygourakis, C.C.; Yoon, S.; Valencia, V.; Boscardin, C.; Moriates, C.; Gonzales, R.; Lawton, M.T. Operating Room Waste: Disposable Supply Utilization in Neurosurgical Procedures. *J. Neurosurg.* **2017**, *126*, 620–625. [CrossRef] [PubMed]
55. Hoenich, N.A.; Levin, R.; Pearce, C. Clinical Waste Generation from Renal Units: Implications and Solutions: CLINICAL WASTE GENERATION FROM RENAL UNITS. *Semin. Dial.* **2005**, *18*, 396–400. [CrossRef] [PubMed]
56. Hendricusdottir, R.; Hussain, A.; Milnthorpe, W.; Bergmann, J.H. Lack of Support in Medical Device Regulation within Academia. *Prosthesis* **2021**, *3*, 1–8. [CrossRef]
57. National Health Service. *The NHS Long Term Plan*; National Health Service: London, UK, 2019.
58. International Organization for Standardization. Environmental Management (ISO Standard Family No. 14000). Available online: https://www.iso.org/iso-14001-environmental-management.html (accessed on 20 September 2021).
59. International Organization for Standardization. Environmentally Conscious Design-Principles, Requirements and Guidance (ISO Standard No. IEC 62430:2019). 2019. Available online: https://www.iso.org/standard/79064.html (accessed on 20 September 2021).
60. Office for Product Safety and Standards; Department of Health and Social Care. Designated Standards: Medical Devices. Available online: https://www.gov.uk/government/publications/designated-standards-medical-devices (accessed on 20 September 2021).
61. Office for Product Safety and Standards; Department of Health and Social Care. Designated Standards: In Vitro Diagnostic Medical Devices. Available online: https://www.gov.uk/government/publications/designated-standards-in-vitro-diagnostic-medical-devices (accessed on 20 September 2021).
62. Office for Product Safety and Standards; Department of Health and Social Care. Designated Standards: Active Implantable Medical Devices. Available online: https://www.gov.uk/government/publications/designated-standards-active-implantable-medical-devices (accessed on 20 September 2021).
63. International Organization for Standardization. Environmental Management Systems-Guidelines for Incorporating Ecodesign (ISO Standard No. 14006:2020). Available online: https://www.iso.org/standard/72644.html (accessed on 20 September 2021).
64. British Standards Institute. Eco-Design. Available online: https://shop.bsigroup.com/Browse-by-Sector/Design/Eco-design/ (accessed on 20 September 2021).
65. Directive 2009/125/EC of the European Parliament and of the Council Establishing a Framework for the Setting of Ecodesign Requirements for Energy-Related Products; p L 285/10-L 285/35. Available online: https://eur-lex.europa.eu/LexUriServ/LexUriServ.do?uri=OJ:L:2009:285:0010:0035:en:PDF (accessed on 20 September 2021).
66. Department for Business, Energy & Industrial Strategy; Office for Product Safety and Standards. Regulations: Ecodesign of Energy-Consuming Products. 2021. Available online: https://www.gov.uk/guidance/placing-energy-related-products-on-the-uk-market (accessed on 20 September 2021).
67. Wagner, W.E. Commons Ignorance: The Failure of Environmental Law to Produce Needed Information on Health and the Environment. *Duke Law J.* **2004**, *53*, 127.
68. Marchant, G.E. The Growing Gap Between Emerging Technologies and the Law. In *The Growing Gap between Emerging Technologies and Legal-Ethical Oversight*; Marchant, G.E., Allenby, B.R., Herkert, J.R., Eds.; The International Library of Ethics, Law and Technology; Springer: Dordrecht, The Netherlands, 2011; Volume 7, pp. 19–33. [CrossRef]
69. Musazzi, U.M.; Marini, V.; Casiraghi, A.; Minghetti, P. Is the European Regulatory Framework Sufficient to Assure the Safety of Citizens Using Health Products Containing Nanomaterials? *Drug Discov. Today* **2017**, *22*, 870–882. [CrossRef] [PubMed]

70. Ren, S.; Li, X.; Yuan, B.; Li, D.; Chen, X. The Effects of Three Types of Environmental Regulation on Eco-Efficiency: A Cross-Region Analysis in China. *J. Clean. Prod.* **2018**, *173*, 245–255. [CrossRef]
71. Medicines & Healthcare products Regulatory Agency. In-House Manufacture of Medical Devices in Great Britain. Available online: https://www.gov.uk/government/publications/in-house-manufacture-of-medical-devices (accessed on 20 September 2021).
72. Medicines & Healthcare products Regulatory Agency. Guidance on Applying Human Factors to Medical Devices. Available online: https://www.gov.uk/government/publications/guidance-on-applying-human-factors-to-medical-devices (accessed on 20 September 2021).
73. Medicines & Healthcare products Regulatory Agency. Devices in Practice: Checklists for Using Medical Devices. Available online: https://www.gov.uk/government/publications/devices-in-practice-checklists-for-using-medical-devices (accessed on 20 September 2021).
74. Renton, D.; Denk, P.; Varban, O. Reprocessed Single-Use Devices in Laparoscopy: Assessment of Cost, Environmental Impact, and Patient Safety. *Surg. Endosc.* **2018**, *32*, 4310–4313. [CrossRef]
75. Eckelman, M.; Mosher, M.; Gonzalez, A.; Sherman, J. Comparative Life Cycle Assessment of Disposable and Reusable Laryngeal Mask Airways. *Anesth. Analg.* **2012**, *114*, 1067–1072. [CrossRef]
76. Heinrichs, H. Sharing Economy: A Potential New Pathway to Sustainability. *GAIA-Ecol. Perspect. Sci. Soc.* **2013**, *22*, 228–231. [CrossRef]
77. Berketa, J.; James, H.; Langlois, N.; Richards, L. A Study of Osseointegrated Dental Implants Following Cremation. *Aust. Dent. J.* **2014**, *59*, 149–155. [CrossRef] [PubMed]
78. Kirkpatrick, J.N.; Ghani, S.N.; Burke, M.C.; Knight, B.P. Postmortem Interrogation and Retrieval of Implantable Pacemakers and Defibrillators: A Survey of Morticians and Patients. *J. Cardiovasc. Electrophysiol.* **2007**, *18*, 478–482. [CrossRef] [PubMed]
79. Laurut, T.; Duran, C.; Pages, A.; Morin, M.-C.; Cavaignac, E. What Is the Cost Burden of Surgical Implant Waste? An Analysis of Surgical Implant Waste in an Orthopedics and Trauma Surgery Department of a French University Hospital in 2016. *Orthop. Traumatol. Surg. Res.* **2019**, *105*, 1205–1209. [CrossRef] [PubMed]
80. Rumble, H. Ashes to Ashes, Rust to Rust? The Recovery and Recycling of Orthopaedic Implants Post-Cremation. In *Residues of Death: Disposal Refigured*; Routledge: Abingdon, UK; New York, NY, USA, 2019; p. 14.

MDPI
St. Alban-Anlage 66
4052 Basel
Switzerland
Tel. +41 61 683 77 34
Fax +41 61 302 89 18
www.mdpi.com

Prosthesis Editorial Office
E-mail: prosthesis@mdpi.com
www.mdpi.com/journal/prosthesis

www.ingramcontent.com/pod-product-compliance
Ingram Content Group UK Ltd.
Pitfield, Milton Keynes, MK11 3LW, UK
UKHW060125010826
14089UKWH00006B/155

9 783036 538938